Aus dem Institut für Mathematik

der Universität zu Lübeck

Direktor:

Prof. Dr. Jürgen Prestin

Non-Linear Registration
Based on Mutual Information

- Theory, Numerics, and Application -

Inauguraldissertation

zur

Erlangung der Doktorwürde der Universität zu Lübeck

- Aus der Technisch-Naturwissenschaftlichen Fakultät -

Vorgelegt von

Dipl.-Inf. Stefan Heldmann

aus Kassel

Lübeck, den 23. Mai 2006

Bibliografische Information der Deutschen Nationalbibliothek

Die Deutsche Nationalbibliothek verzeichnet diese Publikation in der
Deutschen Nationalbibliografie; detaillierte bibliografische Daten sind
im Internet über http://dnb.d-nb.de abrufbar.

ISBN 978-3-8325-1481-5

Logos Verlag Berlin
Comeniushof, Gubener Str. 47,
10243 Berlin
Tel.: +49 030 42 85 10 90
Fax: +49 030 42 85 10 92
INTERNET: http://www.logos-verlag.de

Contents

Preface

Historical Remarks and Aims of the Thesis

In 1896 Wilhelm Conrad Röntgen discovered X-rays which were immediately used for the visualization of internal body regions. This was the hour of birth of radiology and medical imaging. The visualization of inner structures and processes hidden by a surface became a keystone for diagnosis in modern medicine. The final breakthrough was made by Godfrey N. Hounsfield in 1951. Based on an early theoretical work from 1917 of the mathematician Johann Radon, Hounsfield was able to compute three-dimensional images based on a sequence of X-ray images. Thus, he invented the computer tomography (CT).

However, X-ray imaging is not a one-for-all technique. In the last decades many new imaging techniques have been developed, for example magnet resonance imaging (MRI), functional magnet resonance imaging (fMRI), ultra sound (US), positron emission tomography (PET), and single photon emission tomography (SPECT), just to name a few. Each of them is specialized for the visualization of certain features or processes one is interested in. For example, CT is very good for mapping dense structures such as bones, whereas MRI is better suited for visualizing soft-tissue and PET can visualize functional information.

With the development of various imaging techniques and the availability of fast computers the demand for automatic comparison and alignment of images arose. This demand for automatic alignment started a new field in digital imaging - image registration. Image registration is not an easy to solve problem. In particular, the registration of two images that come from different imaging modalities is difficult. In 1995 Viola&Wells and Collignon&Maes et al. independently proposed mutual information (MI) based methods that could deal with this task. Since then, mutual information became very popular, a state-of-the-art tool for image registration, and finally the topic of this thesis. The goal of the thesis is the development and analysis of approaches and numerical methods for non-linear image registration using mutual information. Here, non-linear indicates that we consider non-linear alignments.

Mutual information originates from statistics and information theory. So it is quite natural to carry over concepts and tools from these fields to image registration, as, e.g., Viola&Wells and Collignon&Maes et al. did. Therefore, typical approaches involve the construction of identically independent distributed random variables from images, so-called Parzen window density estimates, and stochastic optimization methods. These methods use random as a key ingredient that makes the theoretical treatment and the comparison of practical results difficult and only allows for a quantitative analysis. Furthermore, efficient optimization schemes, such as Newton's method, will in general perform poorly due to the a lack of a deterministic objective function. To overcome these difficulties, here we consider image registration as a purely deterministic problem. An aim of the thesis is the development of theoretically sound deterministic methods not involving any random settings. Here, in place of statistics and information theory we view mutual information in the light of measure and integration theory.

Image registration may be seen as computing a motion that moves the objects in an image to the "right position" and therefore finally aligns two images. This point of view implies that the "way" matters and is an important part of the registration task. Alternatively, a second interpretation is just to find a transformation that aligns images, i.e., only the outcome matters. Sticking to the first interpretation, we present state-of-the-art partial differential equation (PDE) based approaches for image registration and combine them with the results of our analysis of mutual information. The outcome are parabolic PDEs which evolved over time aligns two images. To this end, we will present semi-implicit numerical schemes. Following the second interpretation, we also present optimization based methods. Here, we are in particular devoted to computational speed. Among others, we consider a limited memory version of the popular BFGS quasi Newton method.

Examining the theoretical aspects of image registration is interesting by itself. Image registration is a highly interdisciplinary topic that connects various mathematical fields such as, e.g., optimization, PDEs, functional analysis, statistics, etc. However, the problem arises from real applications that demand for practical solutions rather than theoretical results. An aim of the thesis is the development of theoretically sound but also applicable methods. In the final chapter of the thesis we demonstrate the performance of the proposed methods in a real application for the registration of 3D PET and CT images.

Outline of the Thesis

The thesis is organized as follows.

Chapter 1: Basic Concepts. In the first chapter we introduce fundamentals and give a general framework for image registration. To this end, we shortly review existing approaches and techniques and finally present the variational approaches for non-linear image registration we are dealing with in this thesis. In particular, we will introduce the modeling of images, how to measure similarity of images with distance measures, and how to measure smoothness of transformations by a so-called smoother. Throughout the thesis, we will consider mutual information as distance measure and three different smoothers yielding the three popular variational approaches for diffusive, curvature and elastic registration.

Chapter 2: Mutual Information. After having introduced the fundamentals of image registration in Chapter 1, here we analyze mutual information with the help of measure and integration theory. The main results of the chapter are the widely unknown fact that mutual information of two images is not well-defined in many cases of practical interest and how to overcome this existence problem by arbitrary small perturbations of the images.

Chapter 3: Practical Treatment of MI. So far we examined theoretical aspects of mutual information. Building on the main results of the previous chapter, we present practical approximation schemes for mutual information. In particular, here we force the interpretation of a deterministic approximation of mutual information rather than the estimation by statistical means. Beside introducing our approximations, a main result of the chapter is revealing the relation of our approach to the common statistical approach that uses kernel density estimation.

Chapter 4: PDE Based Method. In this chapter we turn our attention to PDE based methods to compute a solution of the registration problem. Therefore, first we derive the Euler-Lagrange equations for the diffusive, curvature and elastic registration approach. The Euler-Lagrange equations are non-linear elliptic PDEs that cannot be solved directly. We propose common Gradient-Flow techniques to compute a solution. Subsequently, we turn over to numerical methods. We show how to discretize the PDEs using finite differences and how to compute efficiently mutual information related parts of our scheme using non-equispaced fast Fourier transforms (NFFT). Furthermore, the general numerical method is semi-implicit and therefore requires the solution of large linear systems. In the last part of the chapter, we show how these linear systems can be solved efficiently by fast Fourier techniques (FFT), in particular by discrete sine (DST) and cosine transforms (DCT). We conclude the chapter by discussing the popular multigrid methodology.

Chapter 5: Optimization Based Methods. Alternative to PDE based methods, here we apply optimization techniques to compute a solution to the registration problem. Therefore, in the first step we discretize the registration problem. Subsequently, we apply optimization techniques to derive a numerical method. In particular, we discuss a limited memory BFGS quasi-Newton method.

Chapter 6: 3D PET-CT Registration. In this chapter we demonstrate the performance of the presented registration methods in a real application. In a joint project with the Clinic of Nuclear Medicine of the RWTH Aachen we apply the presented methods for the registration of 3D PET and CT images. In addition to the developed techniques, we also use a multi-level approach. Finally, we present some promising first results which have been manually rated by physicians.

Acknowledgements

I thank my doctoral advisor Prof. Dr. Bernd Fischer for guiding me and his support. In particular, I acknowledge his work in building the SAFIR research group. He provided a friendly and warm atmosphere, opened a lot of opportunities, and, for me, is a beau ideal in academic life.

Furthermore, I'd like to thank my co-referee Prof. Dr. Heinz Handels and Prof. Dr. Thomas Martinez for chairing.

A special thanks goes to PD Dr. Jan Modersitzki. He raised me in the field of image registration since my early days as a student and kept me always "on track" with an uncountable number of discussions. Thank you Jan.

I am grateful for sharing the office and the co-work with Oliver Mahnke. We had many fruitful discussions and a lot of fun. He is responsible for getting me interested in mutual information.

I thank Dr. Hans-Jürgen Kaiser for his co-work in PET-CT registration.

I acknowledge the support from my colleagues at the Institute of Mathematics of the University of Lübeck. In particular I wish to thank Prof. Dr. Jürgen Prestin, Prof. Dr. Daniel Potts, and Prof. Dr. Lutz Mattner.

I am grateful for my doctoral colleagues from the SAFIR research group. We had a great time, a lot of fun, and many inspiring discussions. Here is the whole crew in alphabetical order (ladies first): Silke, Hanno, Martin, Nils, Stefan, Stefan, Sven, and Tia.

Finally, my biggest gratitude goes to my parents Hildegard and Heinrich for their endless love and making everything possible.

Stefan Heldmann, May 2006.

Chapter 1

Basic Concepts

For the enhancement of diagnosis, surgery-planning, etc. one wants to combine different or even complementary information from several imaging techniques. An example is the combination of the high anatomical specificity of a CT image with the functional information of a PET image. Therefore, one overlays the images in a process that is called *image fusion*. An example for the fusion of a CT and a PET image is shown in Figure 1.1. A necessary condition for an image fusion is the spatial alignment of the objects shown in the images. This condition is usually not fulfilled. The images under consideration are taken from different devices at different times. There are global changes of position of the patient specific to each imaging device as well as many inevitable movement artifacts induced by breathing, heartbeat, etc. For a successful image fusion one has to correct these movements and compute an alignment of the images. This process is called *image registration*.

A modern terminology for imaging technique is *imaging modality* or short *modality*. If we aim to register two images of the same modality we speak of *mono-modal image registration* and if the images come from different modalities we speak of *multi-modal image registration*. In this thesis we present and analyze multi-modal image registration methods which utilize a particular statistical concept - mutual information. In the following we introduce the basic terms and a general mathematical framework for image registration. Subsequently, we shortly review state-of-the-art methods and finally present the approach we will use in the latter chapters. We start with defining a mathematical model for images.

1.1 What is an Image?

In medical applications images are generally two-dimensional flat images (e.g. X-ray images), three-dimensional volumes (e.g. CT, MRI), or time series of two and three-dimensional images respectively (e.g. fMRI). Throughout this thesis we model a d-dimensional image I as a function

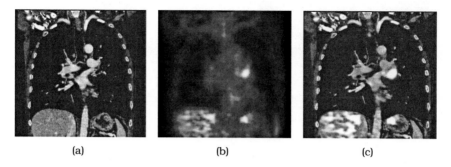

(a)   (b)   (c)

Figure 1.1: (a) CT, (b) PET, (c) fusion

that maps a spatial point $x \in \mathbb{R}^d$ to an *intensity* or *gray value* $I(x) \in \mathbb{R}$.

In the two-dimensional case ($d = 2$) images are said to be composed of so-called *pixels* (*picture elements*) $(x, I(x))$. In three dimensions ($d = 3$) the tuple $(x, I(x))$ is called *voxel* (*volume element*).

Usually, images are rectangular such that we can generally think of an image as a function defined on a cuboid, e.g., $[0, 1]^d$. However, for our latter purposes this plays no major role, so there is no need to be too restrictive. Generally, we consider an image as a compactly supported function.

Definition 1.1 (Image)

Let $\Omega \subset \mathbb{R}^d$ be a domain. A function $I : \mathbb{R}^d \to \mathbb{R}$ is called image, *if*

    a) $\operatorname{supp}(I) \subseteq \overline{\Omega},$ *(compact support)*

    b) $I \in L^1(\Omega),$ *(measurability)*

    c) $\sup\limits_{x \in \Omega} |I(x)| < \infty.$ *(boundedness)*

The space of all images with support in $\overline{\Omega}$ is denoted by $\operatorname{Img}(\Omega)$.

The above definition of an image covers most practical situations. Usually, images are generated by interpolating discrete data. The discrete data is the output of some sensor that measures a small piece of our world. Therefore, the image captures only a cut-out. This is stated by property a). Furthermore, any object one can measure has finite energy. Especially the output of a sensor is a finite measurement of a real-world object. A function that interpolates the observed data should reflect this property. This finiteness is claimed by b) and c).

After we have settled an image model, we now formalize the problem of aligning two images - the registration problem.

1.2 Image Registration

The task of image registration is to compute a geometric alignment of two given images. This is done by transforming the coordinate system of an image. To be more precise, given two images $R, T \in \text{Img}(\Omega)$ the task of the registration is to compute a transformation $\varphi : \mathbb{R}^d \to \mathbb{R}^d$ that deforms T such that R and $T \circ \varphi$ are "similar" on Ω. Since R stays fixed and T gets deformed, R is called *reference image* and T is called *template image*.

A basic necessity for any image registration method is the mathematical definition of similarity of images. When computing an alignment, a registration method optimizes w.r.t. a geometric transformation that aligns the images best. Thus, one has to specify in the registration algorithm what is "best" and therefore what is similarity of images. For the comparison of images we measure similarity by a so-called *distance measure*.

Definition 1.2 (Distance Measure)

A functional $\mathcal{D} : \text{Img}(\Omega) \times \text{Img}(\Omega) \to [-\infty, \infty]$ *is called* distance measure *if*

    a) $\mathcal{D}[R, T] = \mathcal{D}[T, R]$,   *(symmetry)*

    b) $\mathcal{D}[R, R] = \min\limits_{T \in \text{Img}(\Omega)} \mathcal{D}[R, T]$.   *(equality of images)*

First of all, we require symmetry of the distance measure meaning the distance from R to T should be the same as the distance from T to R. To make the idea of a distance measure clear let us fix the argument R for a moment and think of $\mathcal{D}[R, \cdot]$ as a functional only in T. Then we consider an image T to be equal or closest to R if it minimizes the distance measure, i.e., $\mathcal{D}[R, T] = \min\{\mathcal{D}[R, \widehat{T}] : \widehat{T} \in \text{Img}(\Omega)\}$. Clearly, a natural condition for a distance measure is $\mathcal{D}[R, T] = \min$ if $R = T$. Nevertheless, in general there is more than one image T (usually an infinite number) that minimizes $\mathcal{D}[R, \cdot]$ and therefore is equal to R in the sense of $\mathcal{D}$. In particular for multimodal registration this is a wanted property. Thus, a distance measure defines a relation on images. The class of images equal to R in the sense of $\mathcal{D}$ is given by $\{T \in \text{Img}(\Omega) : \mathcal{D}[R, T] = \mathcal{D}[R, R]\}$. Note, this is not an equivalence relation.

Once having established a distance measure, the registration problem reads as follows.

Problem 1.3 (General Registration Problem)

Given two images $R, T \in \mathrm{Img}(\Omega)$ and a distance measure $\mathcal{D}$, find a mapping $\varphi : \mathbb{R}^d \to \mathbb{R}^d$ such that

$$\mathcal{D}[R, T \circ \varphi] \quad \xrightarrow{\varphi} \quad \min .$$

For the sake of generality we have not made any assumptions on φ, yet. The choice of admissible transformations depends on the application and theoretical aspects. We will discuss this point in more detail in section 1.4.

In general, there is no unique transformation φ that solves the registration problem. Due to the above considerations on the distance measure, it might happen that $\mathcal{D}[R, T \circ \varphi_1] = \mathcal{D}[R, T \circ \varphi_2] = \min$ for two distinct transformations $\varphi_1 \neq \varphi_2$. But even if $\mathcal{D}[R, \cdot]$ is minimized by a unique image T (namely $T = R$), there can be an infinite number of solutions φ to the registration problem. This happens for example, if R is invariant w.r.t. a certain class of transformation. Let $R = T$ be a disc and φ_θ a transformation that rotates the coordinate system around the center of the disc by an angle θ. Then $R = T = T \circ \varphi_\theta$ for all angles θ and hence there are infinite many solutions.

Summarizing, the registration problem has in general no unique solution. Speaking in terms of optimization, this means the problem is *ill-posed*. In §1.4 we present state-of-the-art registration techniques addressing this problem by restricting the search space for transformations φ. Before we come to this, first we discuss several classes of distance measures and introduce mutual information.

1.3 Distance Measures

The choice of a suitable distance measure depends on the application. It is a crucial point for image registration that directly affects the computed alignment. In literature various distance measures have been proposed that work in different situations. In particular, the definition of a suitable distance measure for the registration of images that comes from different modalities is not that easy. Here, we use *mutual information (MI)*. It was proposed independently in 1995 by Viola & Wells [62, 63] and Collignon [10] as a similarity measure for multi-modal image registration. Mutual information is a quite general concept originating from information theory. To get a better understanding of mutual information in the context of other distance measures, we give a classification and present two further standard measures. All the functionals in this section are distance measures, i.e., they fulfill Definition 1.2.

Following Roche [48, 47] distance measures can be classified hierarchically. As we will see, mutual information belongs to the most general class.

1.3.1   Intensity Based Distance Measures

Intensity based measures consider the situation of identical images at best. Given an intensity based distance measure $\mathcal{D}$ and two images R and T, then $\mathcal{D}[R, T] = \min$ if and only if $R = T$. Thus, holding R fixed $\mathcal{D}[R, T]$ has a unique global minima at $T = R$.

The most popular candidate for an intensity based distance measure is the so-called *sum of squared differences (SSD)* [5].   For images $R, T \in \text{Img}(\Omega) \cap L^2(\Omega)$, the SSD distance measure $\mathcal{D}^{\text{SSD}}$ is defined as

$$\mathcal{D}^{\text{SSD}}[R, T] := \frac{1}{2} \int_{\Omega} \left(R(x) - T(x) \right)^2 dx. \tag{1.1}$$

The main advantages of the SSD distance measure are that it is easy to interpret and that it has a unique global minima for fixed R. The major drawback is that it cannot be used for multi-modal image registration. Whenever one wants to compare intensities of images directly, the intensity mapping of the imaging devices must be comparable as well. This means we need images from the same modality.

A more general class of distance measures are so-called affine dependency based measures.

1.3.2   Affine Dependency Based Distance Measures

Affine dependency based measures assume their minimum for pairs of images that are identical up to an affine gray value relabeling. Thus, two images $R, T \in \text{Img}(\Omega)$ have minimal distance if an affine function $g(y) = \alpha y + \beta$ with constants $\alpha, \beta \in \mathbb{R}$ exists such that $R = g(T)$ on Ω.

An example for such a distance measure is the so-called *cross-correlation (CC)*. The cross-correlation is similar to the above sum of squared differences, but provides in particular a normalization. Therefore, we define the mean μ and standard deviation σ of a function $f \in L^2(\Omega)$ as

$$\mu[f] = \frac{1}{|\Omega|} \int_{\Omega} f(x) \, dx \quad \text{and} \quad \sigma[f] = \sqrt{\mu \left[(f - \mu[f])^2 \right]}$$

where $|\Omega| := \int_{\Omega} dx$. Then, for images $R, T \in \text{Img}(\Omega) \cap L^2(\Omega)$ the cross-correlation distance measure $\mathcal{D}^{\text{CC}}$ is given by

$$\mathcal{D}^{\text{CC}}[R, T] = - \left| \int_{\Omega} \frac{(R(x) - \mu[R]) \, (T(x) - \mu[T])}{\sigma[R] \, \sigma[T]} \, dx \right|. \tag{1.2}$$

As mentioned above, the cross-correlation and sum of squared differences are strongly related. Rewriting $\mathcal{D}^{\text{SSD}}$ as

$$\mathcal{D}^{\text{SSD}}[R, T] = - \left[\int_{\Omega} R(x) \, T(x) \, dx - \frac{|\Omega|}{2} \left(\mu[R^2] + \mu[T^2] \right) \right]$$

and $\mathcal{D}^{\mathrm{CC}}$ as

$$\sigma[R]\,\sigma[T] \cdot \mathcal{D}^{\mathrm{CC}}[R,T] = -\left|\int_{\Omega} R(x)\,T(x)\,\mathrm{d}x - \mu[R]\,\mu[T]\right|$$

reveals this connection. In particular, restricting the registration to transformations φ with Jacobian $\det(\mathrm{D}\varphi) = 1$ (cf. Definition 2.23 on page 34) and $\mathrm{supp}(T \circ \varphi) \subseteq \Omega$ then $\mu[T \circ \varphi] = \mu[T]$ and $\sigma[T \circ \varphi] = \sigma[T]$. Thus, the mean and the standard deviation do not change under φ and hence are constants. Therefore, the minimization and registration respectively based on $\mathcal{D}^{\mathrm{SSD}}$ coincide with the approach based on $\mathcal{D}^{\mathrm{CC}}$.

In view of multi-modal image registration the situation improves compared to intensity based approaches, because we can classify structural equal images from different modalities, i.e. with distinct gray-value patterns. Nevertheless, the sense of similarity remains too narrow for most cases. The assumption that modalities are related by an affine transformations is too restrictive. Mostly, in practical (multi-modal) applications there is neither an affine or one-to-one, nor a functional dependence at all between the modalities.

Finally, we come to the most general class of distance measures - statistical dependency based measures.

1.3.3 Statistical Dependency Based Distance Measures

Statistical dependency based measures drop the assumption that there is a functional dependence. Nevertheless, in the ideal case two images $R, T \in \mathrm{Img}(\Omega)$ are considered to be most similar if a function $\xi : \mathbb{R} \to \mathbb{R}$ exists such that $R = \xi(T)$ or vice-versa on Ω.

As mentioned above, mutual information is a statistical dependency based distance measure. It is one of the most popular distance measure of all that is still used in mono-modal applications. However, the registration based on MI is the state-of-the-art method in multi-modal applications and the object of this thesis.

Loosely speaking, similarity is measured in the sense how precise one can predict one image from the knowledge of the other. The images are the more similar the better we can predict one image from the other. In statistical terms this means we measure their stochastic dependence. In the next chapter we give a comprehensive analysis. In order to outline the mutual information based registration here, we will forestall both its labeling and properties.

We denote the mutual information of two images R, T by $\mathrm{MI}[R, T]$. Furthermore, $\mathrm{MI}[R, T] \geq 0$ with equality if and only if the images are stochastic independent. For the registration we maximize the mutual information and drive the images as stochastic dependent as possible. Since the registration problem 1.3 is formulated as a minimization problem, we define

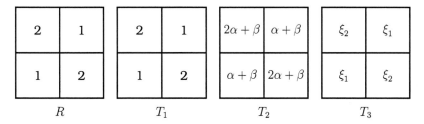

Figure 1.2: T_1 is equal to R w.r.t. an intensity based distance measure; T_2 is equal to R for all α, β w.r.t. an affine dependency based distance measure; T_2 is equal to R for all ξ_1, ξ_2 w.r.t. an statistical dependency based distance measure.

the mutual information distance measure $\mathcal{D}^{\mathrm{MI}}$ as negative mutual information, though

$$\mathcal{D}^{\mathrm{MI}}[R,T] := -\mathrm{MI}[R,T].$$

The registration based on the distance measure $\mathcal{D}^{\mathrm{MI}}$ is called *mutual information based registration*.

Mutual information is a powerful tool and works for a lot of situations. However, it is not easy to interpret as e.g. the sum of squared differences above and its definition, analysis, and numerical treatment is much more involved. But the major difficulty is its generality. As outlined above an image T is considered to be equal to an image R if a mapping ξ exists such that $R = \xi(T)$ and therefore a statistical dependency based measure is minimal. In particular for $\xi = \mathrm{id}$ or an affine mapping such a measure is minimal. Hence, whenever an intensity or affine dependency based measure has a minimum a statistical dependency based measure is minimal, too. Figure 1.2 illustrates this for a small example of 2×2 block images. On one hand the generality of mutual-information is the key for its success in multi-modal registration applications. But on the other hand many solutions exist and we must take care of this when computing a solution of the registration problem.

1.4 Registration Techniques

As outlined in §1.2, the registration problem 1.3 has in general no unique solution. Thus, a direct tackling to compute a solution will not lead to success. Common approaches regularize the problem either by explicitly restricting the search space of admissible transformations or by introducing an additional term to the objective functional. This is not the only reason for the regularization. Usually, there are explicit requirements on the transformations originating from applications. For example, reliable transformations should not produce cracks or foldings or a transformation should be volume preserving [23, 49].

In the following, we present so-called parametric and non-parametric registration techniques. Parametric approaches regularize the problem by restricting the space of admissible transformations. In non-parametric approaches the problem is regularized by introducing an additional term added to the distance measure. Here, we favor the non-parametric approach, because it exhibits more flexibility and allows for a direct modeling of wanted properties for transformations instead of prescribing a particular transformation model.

However, parametric registration techniques are quite popular. For the sake of completeness we next give a rough overview on common parametric approaches. After this we introduce non-parametric methods and present the approaches that are the objectives of this thesis.

1.4.1 Parametric Registration

In parametric registration the admissible transformations are given by an explicit model that depends on parameters. To make ideas clear, let $\Omega \subset \mathbb{R}^2$ and $R, T \in \mathrm{Img}(\Omega)$ be two-dimensional images. Furthermore, we restrict admissible transformations $\varphi : \mathbb{R}^2 \to \mathbb{R}^2$ to translations. Thus φ is given by $\varphi(x) := \varphi_\delta(x) = x + \delta$ with a parameter $\delta \in \mathbb{R}^2$. Then the registration problem 1.3 reads as

$$\mathcal{D}[R, T \circ \varphi_\delta] \quad \xrightarrow{\delta} \quad \min, \quad \delta \in \mathbb{R}^2.$$

Thus, the registration problem simplifies to finding a finite number of real valued parameters instead of computing a whole function. Following examples for parametric registration techniques are frequently used.

Rigid Registration

Rigid registration restricts the search space to rotations and translations. Therefore the transformations $\varphi : \mathbb{R}^d \to \mathbb{R}^d$ are of the form

$$\varphi(x) = Qx + b \qquad \text{with } Q \in \mathbb{R}^{d \times d}, \det(Q) = 1 \text{ and } b \in \mathbb{R}^d.$$

The matrix Q is orthogonal and can be explicitly decomposed into matrices that rotate around unit axes. Therefore, Q can be parameterized by rotation angles.

Affine Registration

Here admissible transformations are linear and allow for rotation, shearing, scaling, and translation. An affine transformation $\varphi : \mathbb{R}^d \to \mathbb{R}^d$ is given by

$$\varphi(x) = Ax + b \qquad \text{with } A \in \mathbb{R}^{d \times d} \text{ and } b \in \mathbb{R}^d.$$

As rigid transformations above, the matrix A can explicitly be decomposed into matrices that depend on geometric parameters for the control of rotation, shearing, and scaling.

Spline Based Registration

Spline based registration allows for non-linear transformations. Therefore admissible transformations φ are explicitly represented as splines. More generally, φ is a linear combination of spline basis functions $\Phi_k :$ $\mathbb{R}^d \to \mathbb{R}^d$, $k = 1, 2, \ldots, N$, though

$$\varphi = \sum_{k=1}^{N} \alpha_k \, \Phi_k \qquad \text{with coefficients } \alpha_k \in \mathbb{R}, \ k = 1, 2, \ldots, N.$$

Common approaches use e.g. tensor products of cubic B-splines or thin plate splines (TPS) as basis functions. The registration is done by optimization w.r.t the coefficients α_k. Thus

$$\mathcal{D}\left[R, T \circ \left(\sum_{k=1}^{N} \alpha_k \, \Phi_k \right)\right] \xrightarrow{\alpha_k} \quad \min.$$

1.4.2   Non-Parametric Registration

In non-parametric approaches the registration problem gets regularized by adding a further term to the distance measure. This term is a functional $\mathcal{S}$ called *smoother*. In general the smoother is a functional that penalizes derivatives of the transformation and therefore enforces a certain smoothness. Moreover, it is convenient to split the transformation $\varphi : \mathbb{R}^d \to \mathbb{R}^d$ into a trivial and a non-trivial part. Therefore we introduce the so-called *displacement* $u : \mathbb{R}^d \to \mathbb{R}^d$ and rewrite the transformation as

$$\varphi = \text{id} - u : \mathbb{R}^d \to \mathbb{R}^d, \qquad x \mapsto \varphi(x) = x - u(x)$$

whereby id is the identity, i.e., $\text{id}(x) = x$. The regularized non-parametric registration problem reads as follows.

Problem 1.4 (Regularized Registration Problem)
Given a pair of images $R, T \in \text{Img}(\Omega)$, a distance measure $\mathcal{D}$, and a smoother $\mathcal{S}$. Find a displacement $u : \mathbb{R}^d \to \mathbb{R}^d$ such that

$$\mathcal{D}[R, T \circ (\text{id} - u)] + \alpha \mathcal{S}[u] \xrightarrow{u} \quad \min, \qquad \alpha > 0.$$

The parameter α is a regularization parameter that weights the similarity of the images versus the smoothness of the transformation. Choosing α

small, the smoother has only little influence and a solution of Problem 1.4 is dominated by minimizing the distance measure. Conversely, choosing α large, much attention is payed to S such that a solution u must be smooth, to minimize the functional.

Alternatively, the parameter α can also be interpreted as a Lagrange multiplier, such that the regularized registration problem is the Lagrangian for the constrained optimization problem

$$\mathcal{D}[R, T \circ (\mathrm{id} - u)] \xrightarrow{u} \min \quad \text{subject to} \quad S[u] \leq c.$$

However, the "smoothness" of a transformation and the "distance" of images are artificial quantities such that it is not clear how to choose values for the parameters α or the constant c. There are heuristics and general methods like the so-called L-curve criterion or general cross-validation (GCV). Nevertheless, even if we apply these methods it is not guaranteed that the result of a registration is what we want. In particular, in medical applications, the evaluation of a registration result is usually made by a visual inspection of the resulting images by a physician. Therefore, here we interpret the parameter α as a model-parameter which has to be chosen manually depending on a concrete application and the image data.

In this thesis we will consider the following three smoothers.

Elastic Smoother

The so-called elastic smoother was discovered for image registration by Broit [4] in 1981. It is physically motivated by the linearized elastic potential given by

$$S^{\mathrm{ELAS}}[u] := \frac{1}{2} \int_\Omega \frac{\mu}{2} \sum_{j,k=1}^d (\partial_j u_k + \partial_k u_j)^2 + \lambda (\mathrm{div}\, u)^2 \, dx \tag{1.3}$$

with the *divergence* $\mathrm{div} = \partial_1 + \partial_2 + \ldots + \partial_d$ and the so-called *Lamé constants* λ and μ that reflect stiffness properties of material. Here we will not discuss the physical motivation. The interested reader might refer to Broit's thesis [4] or to the book by Modersitzki [43].

Diffusive Smoother

The diffusive smoother was proposed by Fischer and Modersitzki in 2002 [18]. There is no physical motivation behind and it was devised for the development of fast and stable algorithms for image registration. The diffusive smoother is given by

$$S^{\mathrm{DIFF}}[u] := \frac{1}{2} \sum_{\ell=1}^d \int_\Omega \|\nabla u_\ell\|^2 \, dx. \tag{1.4}$$

where $\|\cdot\|$ denotes the usual Euclidean norm in $\mathbb{R}^d$ and $\nabla = (\partial_1, \partial_2, \ldots, \partial_d)^\top$ is the gradient. The diffusive smoother penalizes the norm of the gradient of the component functions of the displacement and therefore oscillations.

Curvature Smoother

Another novel smoothing term is given by the so-called curvature smoother. It was introduced by Fischer and Modersitzki in 2002 [17, 19] and is defined as

$$\mathcal{S}^{\mathrm{CURV}}[u] := \frac{1}{2} \sum_{\ell=1}^{d} \int_{\Omega} (\Delta u_\ell)^2 \, \mathrm{d}x, \tag{1.5}$$

where $\Delta = \partial_1^2 + \partial_2^2 + \ldots + \partial_d^2$ is the *Laplace operator*. The curvature smoother can be seen as a rough approximation to the curvature of the displacement. The object of the curvature smoother is not to penalize linear displacements. For an affine transformation $\varphi(x) = Ax + b$ with $A \in \mathbb{R}^{d \times d}$ we have $\mathcal{S}^{\mathrm{CURV}}[\varphi] = 0$ since only second-order derivatives are involved.

In contrast, the elastic and diffusive smoother are only invariant under constants since first-order derivatives are measured.

The elastic, diffusive, and curvature smoothers are quite popular and frequently used for non-parametric registration in the field of medical imaging [4, 12, 18, 17, 19, 23, 22, 24, 26, 28, 29, 30]. Of course, further smoothers have been proposed. For example, in [30] so-called *anisotropic diffusion* originating from computer vision is used as a smoother that, in addition to the displacement field, depends on the template image as well.

However, we will not discuss further approaches here. Throughout this thesis we stick to the elastic, diffusive, and curvature smoother. Consequently, in the following we speak of elastic, diffusive, and curvature registration where we use mutual information as distance measure.

In the following chapters we will analyze and develop numerical methods for the three registration approaches. We start in the following chapter with analyzing the mutual information of images.

Chapter 2

Mutual Information

In the following, we will motivate, define, and analyze mutual information. We will see that we cannot measure the mutual information of arbitrary images. To this end, we will consider in particular so-called *absolutely continuous* images. This class of images partly contains images suitable for the numerical methods presented in the latter chapters. Nevertheless, a considerable amount of images of interest is not covered. The main result in this chapter shows that adding a little noise to the images makes them absolutely continuous and allows for measuring mutual information. The following chapters are based on this result.

Additionally, we will explore the context of mutual information a little. To this end, we consider the class of so-called *discrete images*, e.g., step functions.

We start with shortly outlining needed basic terms from measure theory and integration.

2.1 Basic Terms from Measure Theory

For mutual information we use statistics of intensities of image pairs. These statistics are measurements of sets. Therefore, we introduce the concept of a *σ-algebra*, measures, and integration of functions w.r.t. a certain measure.

Definition 2.1 (σ-Algebra)

Let Ω be a set. A σ-algebra on Ω is a family of subsets $\mathcal{A} \subseteq 2^{\Omega}$ with

a) $\Omega, \emptyset \in \mathcal{A}$,

b) $A, B \in \mathcal{A} \Rightarrow B \setminus A \in \mathcal{A}$,

c) $\bigcup_{n \in \mathbb{N}} A_n \in \mathcal{A}$ *and* $\bigcup_{n \in \mathbb{N}} A_n \in \mathcal{A}$ *for all sequences* $(A_n) \subseteq \mathcal{A}$.

Dealing with sets of a σ-algebra ensures that the intersection, union, and

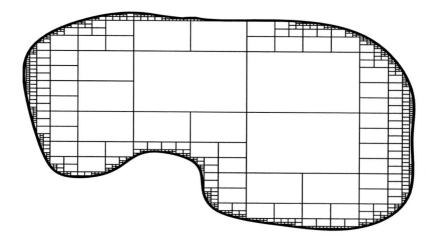

Figure 2.1: Filling of a Borel set with intervals

exclusion is well-defined. In particular, a σ-algebra provides the closure w.r.t. intersection and union. Therefore, we can consider limits of sequences of sets.

A trivial σ-algebra on Ω is its power set 2^Ω. Another - maybe the most important - σ-algebra is the so-called *Borel σ-algebra*.

Definition 2.2 (Borel σ-Algebra)

Let $\mathcal{I}$ denote the set of all left-open intervals in $\mathbb{R}^d$ given by

$$\mathcal{I} := \{]a_1, b_1] \times]a_2, b_2] \times \ldots \times]a_d, b_d] \; : \; a_i < b_i, \; a_i, b_i \in \mathbb{R}, \; i = 1, \ldots, d\}.$$

The Borel σ-algebra $\mathcal{B}(\mathbb{R}^d)$ on $\mathbb{R}^d$ is defined as

$$\mathcal{B}(\mathbb{R}^d) := \bigcap_{\substack{\mathcal{A} \supset \mathcal{I}, \\ \mathcal{A} \text{ is } \sigma\text{-algebra}}} \mathcal{A}.$$

Furthermore, for $\Omega \subset \mathbb{R}^d$ we set $\mathcal{B}(\Omega) := \{\Omega \cap B \; : \; B \in \mathcal{B}(\mathbb{R}^d)\}$.

The above definition of Borel σ-algebra is only one of many. For example, we could use the set of all right-open intervals instead of left-open intervals [15, 31]. The Borel σ-algebra is quite rich, for example $\mathcal{B}(\mathbb{R}^d)$ includes all open sets. However, $\mathcal{B}(\mathbb{R}^d)$ provides the closure of $\mathcal{I}$ w.r.t. intersection, union, and exclusion. Therefore, any set that can be filled with intervals is a Borel set. An illustration for that is given in Figure 2.1. Furthermore, for countable $\Omega = \{\omega_1, \omega_2, \omega_3, \ldots\}$ the Borel σ-algebra on Ω is the power set, that is $\mathcal{B}(\Omega) = 2^\Omega$.

Now we come to measuring sets.

Definition 2.3 (Measure, Probability Measure)

Let $\mathcal{A}$ be a σ-algebra on Ω. A measure *is a mapping $\mu : \mathcal{A} \to [0, \infty]$ with*

$$\mu(\emptyset) = 0 \quad \text{and} \quad \mu\left(\bigcup_{n \in \mathbb{N}} A_n\right) = \sum_{n \in \mathbb{N}} \mu(A_n) \text{ where } A_i \cap A_j = \emptyset \text{ for } i \neq j.$$

Furthermore, $(\Omega, \mathcal{A}, \mu)$ is called measure space. *If $\mu(\Omega) = 1$ then μ is called* probability measure *and $(\Omega, \mathcal{A}, \mu)$* probability space.

Two important examples for measures are the *Lebesgue* and *counting measure*.

Let $I =]a_1, b_1] \times]a_2, b_2] \times \ldots \times]a_d, b_d]$ be a left-open interval in the set $\mathcal{I}$ as in Definition 2.2. We define the *volume* of I as

$$\text{vol}(I) := (b_1 - a_1)(b_2 - a_2) \cdots (b_d - a_d).$$

Referring to [2, 31], the d-dimensional *Lebesgue measure* $\mathcal{L}^d$ is defined as the unique measure on $\mathcal{B}(\mathbb{R}^d)$ with

$$\mathcal{L}^d(I) = \text{vol}(I) \quad \text{for all intervals } I \in \mathcal{I}. \tag{2.1}$$

For a countable set $\Omega = \{\omega_1, \omega_2, \omega_3, \ldots\} \subset \mathbb{R}^d$ the *counting measure* on $\mathcal{B}(\Omega) = 2^\Omega$ is given by

$$\#^d(A) := \sum_{\omega \in A} \quad \text{for all } A \in 2^\Omega. \tag{2.2}$$

In the following, we still omit the superscript indicating the dimensions, i.e., we write shortly $\#$ instead of $\#^d$.

Furthermore, the d-dimensional Lebesgue and counting measure, respectively, are examples of so-called *product measures*. In general, given two measure spaces $(\Omega, \mathcal{A}, \mu)$ and $(\Omega', \mathcal{B}, \nu)$ the product measure $\mu \otimes \nu$ of μ and ν is defined by

$$\mu \otimes \nu(A \times B) := \mu(A)\mu(B) \quad \text{for all } A \in \mathcal{A}, \, B \in \mathcal{B}. \tag{2.3}$$

For example $\mathcal{L}^d$ is the d-times product of the one-dimensional Lebesgue measure $\mathcal{L}$, i.e.,

$$\mathcal{L}^d = \underbrace{\mathcal{L} \otimes \mathcal{L} \otimes \ldots \otimes \mathcal{L}}_{d\text{-times}}.$$

Next, we turn to integrating real-valued functions w.r.t. a certain measure. Before we define the integral, we need a further term - measurability. Let $f : \Omega \to \Omega'$, $\mathcal{A}$ be a σ-algebra on Ω and $\mathcal{A}'$ be a σ-algebra on Ω'. Then f is called *measurable* if

$$f^{-1}(A') \in \mathcal{A} \quad \text{for all } A' \in \mathcal{A}'.$$

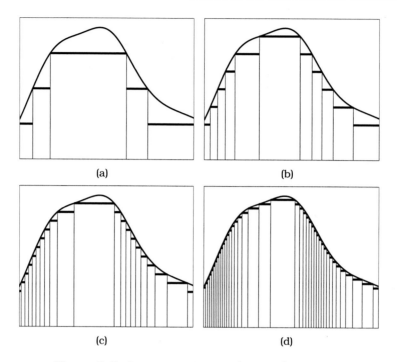

Figure 2.2: Approximation with step functions

If nothing else is said and $\mathcal{A}$ and $\mathcal{A}'$, respectively are not specified, we always consider the Borel σ-algebra on Ω and Ω', respectively, i.e., $\mathcal{A} = \mathcal{B}(\Omega)$ and $\mathcal{A}' = \mathcal{B}(\Omega')$.

Definition 2.4 (μ-Integral, μ-Integrable)

Let $(\Omega, \mathcal{A}, \mu)$ be a measure space and $f : \Omega \to [0, \infty]$ measurable. The μ-integral of f is defined as

$$\int_\Omega f(\omega)\, d\mu(\omega) := \lim_{n \to \infty} 2^{-n} \sum_{j=1}^{\infty} \mu(\{\omega \in \Omega : f(\omega) > j2^{-n}\}).$$

For measurable $f : \Omega \to \overline{\mathbb{R}}$ we set $f^+ := \max\{0, f\}$, $f^- := \min\{0, -f\}$ and $|f| = f^+ + f^-$. The μ-integral of f is defined as

$$\int_\Omega f(\omega)\, d\mu(\omega) := \int_\Omega f^+(\omega)\, d\mu(\omega) - \int_\Omega f^-(\omega)\, d\mu(\omega).$$

If $\int_\Omega |f(\omega)|\, d\mu(\omega) < \infty$, we say f is μ-integrable.

For a short notation, in the following we sometimes omit the integration variable, i.e.,

$$\int_\Omega f\, d\mu := \int_\Omega f(\omega) d\mu(\omega).$$

Furthermore, when we integrate w.r.t. the Lebesgue measure (2.1) we also write as common dx instead of $\mathcal{L}^d$, i.e.,

$$\int_\Omega f(x)\,dx := \int_\Omega f(x)\,d\mathcal{L}^d(x) \quad\text{and}\quad \int_\Omega f\,dx := \int_\Omega f\,d\mathcal{L}^d.$$

To make the idea of the μ-integral clear, let us consider the integral of step functions. In general, a step function $s : \Omega \subset \mathbb{R}^d \to \overline{\mathbb{R}}$ reads

$$s(x) = \sum_{j=1}^N y_j\,1_{S_j}(x) \quad\text{with } y_j \in \overline{\mathbb{R}},\ S_j \in \mathcal{B}(\Omega),\ N \in \mathbb{N},$$

and indicator functions

$$1_{S_j}(x) := \begin{cases} 1 & \text{if } x \in S_j, \\ 0 & \text{if } x \notin S_j. \end{cases}$$

The integral of a step function is given by $\int_\Omega s(x)\,d\mu(x) = \sum_{j=1}^N y_j\,\mu(S_j)$ [15, §4]. Keeping this in mind, let f be a non-negative, measurable function. Furthermore, to keep things simple let f additionally be bounded from above by some constant $M < \infty$. Then

$$\int_\Omega f(\omega)\,d\mu(\omega) = \lim_{n\to\infty} 2^{-n} \sum_{j=1}^{\lfloor M2^n \rfloor} \mu(\{\omega \in \Omega : f(\omega) > j2^{-n}\})$$

$$= \lim_{n\to\infty} \sum_{j=1}^{\lfloor M2^n \rfloor} j2^{-n}\mu(\{\omega \in \Omega : j2^{-n} < f(\omega) \le (j+1)2^{-n}\})$$

$$= \lim_{n\to\infty} \int_\Omega s_n(x)\,d\mu(x)$$

with step functions

$$s_n(x) := \sum_{j=1}^{\lfloor M2^n \rfloor} y_j^n\,1_{S_j^n}(x) \quad\text{where } y_j^n := j/2^n \text{ and } S_j^n := f^{-1}(]j/2^n, (j+1)/2^n]).$$

Thus, we approximate f by step functions s_n and the integral over f reveals as the limit of the integral over s_n (cf. Figure 2.2).

Furthermore, let Ω be a countable set and $(\Omega, 2^\Omega, \#)$ the measure space with the counting measure $\#$ as in (2.2). Then the $\#$-integral of a measurable function $f : \Omega \to \overline{\mathbb{R}}$ is a sum, given by

$$\int_\Omega f(\omega)\,d\#(\omega) = \sum_{\omega \in \Omega} f(\omega).$$

Now we have all the terms we need in the following. Next, we motivate the use of mutual information as a distance measure for images. Mutual information is a measure for stochastic dependence. Therefore, we consider the meaning of stochastic dependence for images.

2.2 Stochastic Dependence of Images

Let R and T be two images and $R(x) = r$ at a certain point x. What can we state about the intensity of the second image with $T(x) = t$? In the best case there is a functional dependence between the images. That is, a function ξ exists, such that $\xi(R) = T$ and therefore $\xi(r) = t$. In general, such a functional dependence is missing, but nevertheless the relation between the intensities is not completely random. In contrast, the worst case is that we cannot state anything about the intensity t. That is, the images are stochastically independent.

The idea of mutual information is to measure the stochastic dependence of images. This is done in terms of frequencies and probabilities respectively of co-occurrences of gray values.

To be more concrete let $R, T : \Omega \to Y \in \operatorname{Img}(\Omega)$ and $A, B \subseteq Y$ two sets of gray values. We say gray values $r \in A$ and $t \in B$ co-occur if there exists a point $x \in \Omega$ such that $R(x) = r$ and $T(x) = t$. Then the set of points with co-occurring gray values in A and B is given by the preimage of $A \times B$ of the joint function $(R, T) : \Omega \to Y^2$, i.e.,

$$(R, T)^{-1}(A \times B) = \{x \in \Omega : R(x) \in A \text{ and } T(x) \in B\}.$$

The frequency of co-occurring gray values is given by "counting" the elements of the preimage. Therefore, we measure the preimage with a probability measure. This probability measure will be nothing else than a normalized version of the d-dimensional Lebesgue measure.

Definition 2.5 (Normalized Lebesgue Measure)
Let $\Omega \subset \mathbb{R}^d$. The normalized Lebesgue measure $P : \mathcal{B}(\mathbb{R}^d) \to [0, 1]$ on Ω is given by

$$P(A) := \frac{1}{|\Omega|} \int_\Omega 1_A(x) \, dx \quad \text{for all } A \in \mathcal{B}(\mathbb{R}^d),$$

with $|\Omega| := \int_\Omega dx = \mathcal{L}^d(\Omega)$.

Then the probability that R takes values in A while T simultaneously takes values in B is

$$P((R, T)^{-1}(A \times B)) = \frac{1}{|\Omega|} \int_\Omega 1_{\{x \in \Omega : R(x) \in A \text{ and } T(x) \in B\}}(y) \, dy.$$

The function $P((R, T)^{-1}(\cdot))$ is the so-called *joint-distribution* of the images R and T. It is central in the computation of mutual information.

$$R \qquad\qquad T \qquad\qquad \mathrm{hist}(R,T)$$

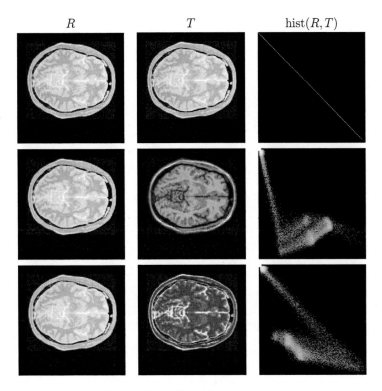

Figure 2.3: Joint histograms

Definition 2.6 (Distributions)

Let $Y \subseteq \mathbb{R}^k$, and P be the normalized Lebesgue measure on Ω. The distribution $P_f : \mathcal{B}(Y) \to [0,1]$ of a measurable function $f : \Omega \to Y$ is given by

$$P_f(A) := P(f^{-1}(A)), \quad A \in \mathcal{B}(Y)$$

where $f^{-1}(A)$ is the preimage of A, i.e., $f^{-1}(A) = \{\omega \in \Omega \ : \ f(\omega) \in A\}$. In particular, for images $R, T : \Omega \to Y \in \mathrm{Img}(\Omega)$ the distribution

$$P_{RT}(A \times B) = P((R,T)^{-1}(A \times B)), \quad A \times B \in \mathcal{B}(Y \times Y).$$

of the joint function $(R,T) : \mathbb{R}^d \to Y^2$ is called joint distribution and the distributions

$$P_R(A) = P(R^{-1}(A)) \quad \text{and} \quad P_T(B) = P(T^{-1}(B)), \quad A, B \in \mathcal{B}(Y)$$

are called marginal distributions.

Example 2.7

Let $R, T \in \text{Img}(\Omega)$ and $B_k :=]k\beta - \frac{\beta}{2}, k\beta + \frac{\beta}{2}]$, $k \in \mathbb{Z}$ be so-called bins with bin-size $\beta > 0$. The joint histogram $\text{hist}(R, T)$ is a matrix defined as the binning of the joint distribution P_{RT} given by

$$\text{hist}(R, T) := \left(P_{RT}(B_k \times B_\ell) \right)_{k, \ell \in \mathbb{Z}}.$$

Figure 2.3 gives a visual impression for the joint histogram and therefore the joint distribution of images.

Two images $R, T : \Omega \to Y \in \text{Img}(\Omega)$ are said to be stochastically independent if the joint distribution is the product measure of its marginal distributions, i.e., $P_{RT} = P_R \otimes P_T$ and therefore

$$P_{RT}(A \times B) = P_R(A)\, P_T(B) \quad \text{for all } A, B \in \mathcal{B}(Y). \tag{2.4}$$

For a further understanding of stochastic dependence, let us consider the so-called *conditional probability distribution* of images. It is defined as

$$P_{R|T}(A|B) := \frac{P_{RT}(A \times B)}{P_T(B)} \quad \text{provided } P_T(B) > 0. \tag{2.5}$$

Then $P_{R|T}(A|B)$ gives the relative probability that R takes values in A while already knowing T takes values in B. Therefore, the conditional probability tells us how well we can predict R from the knowledge of T. In the case of stochastic independent images we have

$$P_{R|T}(A|B) = \frac{P_{RT}(A \times B)}{P_T(B)} = \frac{P_R(A)P_T(B)}{P_T(B)} = P_R(A).$$

Thus, the conditional probability for $R(x) \in A$ is not influenced by the information $T(x) \in B$. In other words, the knowledge $T(x) \in B$ does not help to make any statement on the probability for $R(x)$ is an intensity in the set A. Therefore, the information $T(x) \in B$ is useless to predict $R(x)$. In contrast, assume a one-to-one mapping $\xi : \mathbb{R} \to \mathbb{R}$ exists such that $\xi(R(x)) = T(x)$ for all $x \in \Omega$. This yields

$$P_{RT}(\xi^{-1}(A) \times A) = \frac{1}{|\Omega|} \int_\Omega 1_{\{x \in \Omega : R(x) \in \xi^{-1}(A) \text{ and } T(x) \in A\}}(y)\, dy$$

$$= \frac{1}{|\Omega|} \int_\Omega 1_{\{x \in \Omega : \xi(R(x)) \in A \text{ and } T(x) \in A\}}(y)\, dy$$

$$= \frac{1}{|\Omega|} \int_\Omega 1_{\{x \in \Omega : T(x) \in A\}}(y)\, dy$$

$$= P_T(A).$$

Therefore, $P_{R|T}(\xi^{-1}(A)|A) = 1$ such that from the knowledge $T(x) = t$ we can predict $R(x) = \xi^{-1}(t)$ with probability one.

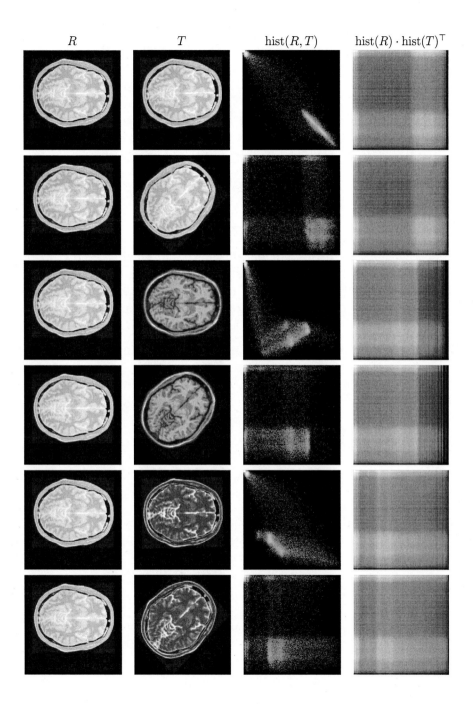

Figure 2.4: Histograms of rotated images by 1 and 45 degrees

Example 2.8

*Let $R, T \in \mathrm{Img}(\Omega)$ and $B_k :=]k\beta - \frac{\beta}{2}, k\beta + \frac{\beta}{2}]$, $k \in \mathbb{Z}$ be bins with bin-size $\beta > 0$
as in Example 2.7. Analogous to the joint histogram $\mathrm{hist}(R, T)$, we define
the marginal histograms as the binning of the marginal distributions, i.e.,*

$$\mathrm{hist}(R) := \Big(P_R(B_k)\Big)_{k \in \mathbb{Z}} \quad \text{and} \quad \mathrm{hist}(T) := \Big(P_T(B_\ell)\Big)_{\ell \in \mathbb{Z}}.$$

Thus, for the binning of the product measure $P_R \otimes P_T$ we have

$$\Big(P_R \otimes P_T(B_k \times B_\ell)\Big)_{k,\ell \in \mathbb{Z}} = \Big(P_R(B_k)P_R(B_\ell)\Big)_{k,\ell \in \mathbb{Z}} = \mathrm{hist}(R)\,\mathrm{hist}(T)^\top$$

*Furthermore, for stochastic independent images we have $P_{RT} = P_R \otimes P_T$
yielding*

$$\mathrm{hist}(R, T) = \mathrm{hist}(R)\,\mathrm{hist}(T)^\top \quad \text{provided } R, T \text{ stoch. independent.}$$

*Figure 2.4 visualizes the joint distribution P_{RT} and the product of the
marginals $P_R \otimes P_T$ for rotated versions of the images given in Figure 2.3.
We can see the joint histograms of the unrotated images get dispersed
and the joint histogram gets closer to the product of its marginals. This
gives an impression for the stochastic dependence of the images.*

As already mentioned, the mutual information measures the stochastic
dependence of the images. It is also called a measure of uncertainty in
the sense how well one can predict one image from the other. Now, we
come to its definition.

2.3 Definition of MI

For the definition of the MI we need a further description of distributions
- densities.

Definition 2.9 (Absolutely μ-Continuous, μ-Density)

*Let $f : \Omega \to Y \subset \mathbb{R}^k$ with distribution P_f and $(Y, \mathcal{B}(Y), \mu)$ be a measure
space. If $p_f \in L^1(Y)$, non-negative exists with*

$$P_f(A) = \int_A p_f(x)\,\mathrm{d}\mu(x) \quad \text{for all } A \in \mathcal{B}(Y)$$

*then p_f is called μ-density and we say f and P_f respectively are abso-
lutely μ-continuous. The space of absolutely μ-continuous functions is
denoted by*

$$AC(\Omega; \mu) := \{f : \Omega \to Y : f \text{ is absolutely } \mu\text{-continuous}\}.$$

*In particular, for $\mu = \mathcal{L}^k$, to make it short, p_f is called density and we
say f and P_f respectively are absolutely continuous.*

The existence of a μ-density states that we can compute a distribution w.r.t. the measure μ. The crucial point is that we cannot find a μ-density for every measure μ. Next, we give some examples for densities, their existence, and non-existence w.r.t. a certain measure.

Example 2.10

a) *The normalized Lebesgue measure P on Ω from Definition 2.5 is absolutely continuous with density $p = \frac{1}{|\Omega|}1_\Omega$, since*

$$P(A) = \frac{1}{|\Omega|}\int_\Omega 1_A(x)\ dx = \frac{1}{|\Omega|}\int_A 1_\Omega(x)\ dx = \int_A p(x)\ dx$$

for all $A \in \mathcal{B}(\mathbb{R}^k)$.

b) *Let $f : [0,1] \rightarrow \{0,1\}$ be a function with*

$$f(x) = \begin{cases} 0, & 0 \le x < \frac{1}{4}\ \text{and} \\ 1, & \frac{1}{4} \le x \le 1. \end{cases}$$

Therefore, the distribution P_f of f is given by

$$P_f(\{0\}) = \frac{1}{4} \quad \text{and} \quad P_f(\{1\}) = \frac{3}{4}.$$

Then f is absolutely #-continuous with #-density $p_f : \{0,1\} \rightarrow \mathbb{R}_+$ given by $p_f(0) = P_f(\{0\}) = \frac{1}{4}$ and $p_f(1) = P_f(\{1\}) = \frac{3}{4}$, since

$$P_f(\{0\}) = \int_{\{0\}} p_f\ d\# = \sum_{a \in \{0\}} p_f(a) = \frac{1}{4},$$

$$P_f(\{1\}) = \int_{\{1\}} p_f\ d\# = \sum_{a \in \{1\}} p_f(a) = \frac{3}{4},$$

$$P_f(\{0,1\}) = \int_{\{0,1\}} p_f\ d\# = \sum_{a \in \{0,1\}} p_f(a) = \frac{1}{4} + \frac{3}{4} = 1.$$

c) *The function f from b) is not absolutely continuous. For example*

$$P_f(\{0\}) = \frac{1}{4} > 0 \quad \text{but} \quad \int_{\{0\}} \varphi\ dx = 0\ \text{for all}\ \varphi \in L^1(\mathbb{R}),$$

since $\mathcal{L}(\{0\}) = 0$. Hence, no function $p_f \in L^1(\mathbb{R})$ exists which integrates to $\frac{1}{4} = P_f(\{0\})$ on $\{0\}$.

The densities of the distributions P_{RT}, P_R, and P_T are central for us. To this end, we give an extra definition.

Definition 2.11 (Joint Density and Marginals)

Let $R, T \in \mathrm{Img}(\Omega)$ be absolutely μ-continuous images with μ-densities p_R and p_T, respectively. Furthermore, let $(R, T) \in AC(\Omega; \mu^2)$ with μ^2-density p_{RT}. We call p_{RT} with

$$P_{RT}(A \times B) = \int_{A \times B} p_{RT}(r, t) \, \mathrm{d}\mu^2(r, t)$$

joint density and p_R, p_T with

$$P_R(A) = \int_A p_R(r) \, \mathrm{d}\mu(r) \quad \text{and} \quad P_R(B) = \int_B p_T(t) \, \mathrm{d}\mu(t)$$

are called marginals.

In the above definition, we required the images R and T to be absolutely μ-continuous and further the joint images (R, T) to be absolutely μ^2-continuous. It turns out that the marginals can be computed from the joint density if it exists. Therefore, the absolute μ^2-continuity of (R, T) implies the absolute μ-continuity of R and T. This well-known fact is stated by the following Lemma.

Lemma 2.12 (Computation of Marginals)

Let $(R, T) : \Omega \to Y^2 \in \mathrm{Img}(\Omega)^2 \cap AC(\Omega; \mu^2)$ with joint distribution P_{RT} and joint μ^2-density p_{RT}. Then $R, T \in AC(\Omega; \mu)$ with μ-densities p_R, p_T given by

$$p_R(r) = \int_Y p_{RT}(r, t) \, \mathrm{d}\mu(t) \quad \text{and} \quad p_T(t) = \int_Y p_{RT}(r, t) \, \mathrm{d}\mu(r),$$

and marginal distributions

$$P_R(A) = P_{RT}(A \times Y) \quad \text{and} \quad P_T(B) = P_{RT}(Y \times B).$$

Proof. Since $R, T : \Omega \to Y$ we have for all $A \subseteq Y$

$$\begin{aligned}
R^{-1}(A) &= \{x \in \Omega \,:\, R(x) \in A\} \\
&= \{x \in \Omega \,:\, R(x) \in A \text{ and } T(x) \in Y\} \\
&= (R, T)^{-1}(A \times Y).
\end{aligned}$$

Hence $P_R(A) = P_{RT}(A \times Y)$ and therefore

$$P_R(A) = \int_{A \times Y} p_{RT}(r, t) \, \mathrm{d}\mu^2(r, t) = \int_A \underbrace{\int_Y p_{RT}(r, t) \, \mathrm{d}\mu(t)}_{=p_R(r)} \, \mathrm{d}\mu(r)$$

Thus, R is absolutely μ-continuous. The proof for T is along the same line. ∎

Now we are in a position to give a definition of mutual information.

Definition 2.13 (Mutual μ-Information)

Let $R, T \in \text{Img}(\Omega)$ be images with ranges Y, and $(R,T) \in AC(\Omega; \mu^2)$ with joint μ^2-density p_{RT} and marginals p_R, p_T. The mutual μ-information (MI) is defined as

$$\text{MI}[R, T] = \int_{Y \times Y} p_{RT}(r, t) \, \log \frac{p_{RT}(r, t)}{p_R(r) \, p_T(t)} \, \mathrm{d}\mu(r)\mathrm{d}\mu(t)$$

whereby $\log \frac{p_{RT}(r,t)}{p_R(r) \, p_T(t)} := 0$ if $p_{RT}(r, t) = 0$.

The MI compares distributions by their μ-densities. Therefore, the measure μ can be interpreted as a "reference measure". The crucial point is that no measure μ exists, for which all image pairs are absolutely μ^2-continuous. This is still the reason for our quite abstract definition of densities and the mutual information.

Next, we give two further representations for the MI that are used in common - MI as *Kullback-Leibler divergence* and as sum of *entropies*.

2.3.1 MI as Kullback-Leibler Divergence

Here we focus on the interpretation of MI as measure for the stochastic dependence of images. Therefore, we measure the distance of a joint distribution to the product of its marginals. This can be expressed by the so-called Kullback-Leibler divergence.

Definition 2.14 (Kullback-Leibler μ-Divergence)

Let P, Q be two absolutely μ-continuous probability measures with μ-densities p and q. The Kullback-Leibler μ-divergence of P and Q is defined as

$$D(P||Q) := \int_{\Omega} p(x) \, \log \frac{p(x)}{q(x)} \, \mathrm{d}\mu(x).$$

The Kullback-Leibler divergence measures the distance between probability measures and therefore distributions. When we look at the integral in Definition 2.14 substituting p with a joint density p_{RT} and q with the product of its marginals $p_R p_T$ yields the MI integral. By its definition p_{RT} is the density of a joint distribution P_{RT}. Furthermore, as we will show in the proof of the following Lemma, the product $p_R p_T$ is the density of $P_R \otimes P_T$.

Lemma 2.15 (MI as Kullback-Leibler Divergence)
Let $(R,T) \in \mathrm{Img}(\Omega)^2 \cap AC(\Omega; \mu^2)$. Then $\mathrm{MI}[R,T]$ is given by

$$\mathrm{MI}[R,T] = D(P_{RT} \| P_R \otimes P_T).$$

Proof. For $P_R \otimes P_T$ holds

$$(P_R \otimes P_T)(A \times B) = \int_A p_R(r) \, d\mu(r) \int_B p_T(t) \, d\mu(t)$$

$$= \int_A \int_B p_R(r) p_T(t) \, d\mu(r) d\mu(t)$$

$$= \int_{A \times B} p_R(r) p_T(t) \, d\mu^2(r,t).$$

Thus, $P_R \otimes P_T$ is absolutely μ^2-continuous with μ^2-density $p_R p_T$. Then the Lemma follows from the definition of the Kullback-Leibler divergence. ∎

Thus, the MI reveals as the distance of the joint distribution to the product of its marginal distributions. Since $P_R \otimes P_T$ equals P_{RT} for stochastically independent images (cf. 2.4) the MI measures the distance of the true joint distribution P_{RT} to the imaginary joint distribution $P_R \otimes P_T$ of stochastic independent images. Note that this measurement is totally data driven. We do not measure the distance of P_{RT} to another fixed distribution that is independent of the images but to $P_R \otimes P_T$ that still depends on the images as well.

2.3.2 MI in Terms of Entropy

Another formulation of the mutual information that is often used is given in terms of so-called entropies.

Definition 2.16 (μ-Entropy)
Let $Y \subseteq \mathbb{R}^k$ and $f : \Omega \to Y \in AC(\Omega; \mu)$ with μ-density p_f. The μ-entropy of f is defined as

$$H[f] := - \int_Y p_f(y) \log p_f(y) \, d\mu(y)$$

with the convention $\log(0) := 0$.

Entropy originates from information theory and coding where μ is commonly the counting measure $\#$ such that the μ-integral is a sum. Originally, the entropy is motivated as follows:
Whenever one wants to send messages over a digital channel one has to code the messages. Therefore, digits from an alphabet are used. An example for an alphabet is the ASCII table where each letter is coded with eight bits. Then p_f gives the probability that an element of the alphabet

is used. The entropy can be interpreted as the expectation for the code-length of a message when using a certain alphabet. For further reading on the information theoretic background of mutual information refer, e.g., to the books [11, 41].

Lemma 2.17 (MI in Terms of Entropy)

Let $(R, T) \in \text{Img}(\Omega)^2 \cap AC(\Omega; \mu^2)$. Then $\text{MI}[R, T]$ is given by

$$\text{MI}[R, T] = H[R] + H[T] - H[(R, T)].$$

Proof. Let $Y \subseteq \mathbb{R}$ be the range of both R and T. Then expanding the log-term in the definition of the MI yields

$$\text{MI}[R, T] = \int_{Y \times Y} p_{RT} \log \frac{p_{RT}}{p_R p_T} \, d\mu^2$$

$$= \underbrace{\int_{Y \times Y} p_{RT} \log p_{RT} \, d\mu^2}_{= -H[(R,T)]} - \int_{Y \times Y} p_{RT} \log p_R \, d\mu^2 - \int_{Y \times Y} p_{RT} \log p_T \, d\mu^2.$$

Furthermore,

$$-\int_{Y \times Y} p_{RT} \log p_R \, d\mu^2 = -\int_Y \underbrace{\left(\int_Y p_{RT} \, d\mu \right)}_{= p_R \text{ (cf. Lemma 2.12)}} \log p_R \, d\mu = H[R]$$

and analogously $-\int_{Y \times Y} p_{RT} \log p_R \, d\mu^2 = H[T]$.   ∎

2.4 Properties of MI

For the registration problem we use mutual information to measure the distance of images. The following theorem gives some properties that justify its use as a distance measure.

Theorem 2.18 (Properties of MI)

Let $(R, T) \in \text{Img}(\Omega)^2 \cap AC(\Omega; \mu^2)$. Then the following statements hold:

a) $\text{MI}[R, T] = \text{MI}[T, R]$ (symmetry)

b) $\text{MI}[R, T] \geq 0$ with equality if and only if R and T are stochastically independent.   (positive definiteness)

Proof. a) Follows directly from the definition of the joint distribution and the mutual information.

b) Let $Y \subseteq \mathbb{R}$ be both the range of R and T. Since $\log x \leq x - 1$ for all $x \geq 0$ we have

$$x \log y = x \log \frac{y}{x} + x \log x \leq y - x + x \log x$$

with equality if and only if $x = y$. Thus, setting $x = p_{RT}$ and $y = p_R p_T$ yields

$$\int_{Y \times Y} p_{RT} \log \frac{p_{RT}}{p_R p_T} \, \mathrm{d}\mu^2 = \int_{Y \times Y} p_{RT} \log p_{RT} \, \mathrm{d}\mu^2 - \int_{Y \times Y} p_{RT} \log p_R p_T \, \mathrm{d}\mu^2$$

$$\geq \int_{Y \times Y} p_{RT} \log p_{RT} \, \mathrm{d}\mu^2$$

$$- \underbrace{\int_{Y \times Y} p_R p_T \, \mathrm{d}\mu^2}_{=1} + \underbrace{\int_{Y \times Y} p_{RT} \, \mathrm{d}\mu^2}_{=1} - \int_{Y \times Y} p_{RT} \log p_{RT} \, \mathrm{d}\mu^2$$

$$=0$$

with equality if and only if $p_{RT} = p_R p_T$. ■

Furthermore, we give another important representation of the MI that we will use later for a variational formulation of the registration problem.

Theorem 2.19 (MI as Integral on Ω)

Let $(R, T) \in \mathrm{Img}(\Omega)^2 \cap AC(\Omega; \mu^2)$. Then $\mathrm{MI}[R, T]$ is given by

$$\mathrm{MI}[R, T] = \frac{1}{|\Omega|} \int_\Omega \log \frac{p_{RT}(R(x), T(x))}{p_R(R(x)) \, p_T(T(x))} \, \mathrm{d}x.$$

Proof. Let $Y \subseteq \mathbb{R}$ be both the range of R and T and $g := \log \frac{p_{RT}}{p_R p_T}$. Then

$$\mathrm{MI}[R, T] = \int_{Y \times Y} p_{RT}(r, t) \, g(r, t) \, \mathrm{d}(r, t) = \int_{Y \times Y} g(r, t) \, \mathrm{d}P_{RT}(r, t)$$

and applying the general transformation rule (c.f. Theorem A.5) yields

$$\mathrm{MI}[R, T] = \int_{Y \times Y} g(r, t) \, \mathrm{d}P_{RT}(r, t)$$

$$= \int_\Omega g(R(x), T(x)) \, \mathrm{d}P(x)$$

$$= \frac{1}{|\Omega|} \int_\Omega g(R(x), T(x)) \, \mathrm{d}x.$$

■

This representation is remarkable, since we no longer integrate over the intensities but the spatial support of the images. Furthermore, we have decoupled μ-densities with the μ-integral. Hence, we only need that images are absolutely μ-continuous to *any* measure μ whereas the integration is done w.r.t. the Lebesgue measure.

In the previous chapter we presented a hierarchy of distance measures and explicitly introduced the sum of squared differences and cross correlation (cf. (1.1), (1.2) on page 5) as representatives for intensity based and affine dependency based measures respectively. Using the terms and tools introduced in this chapter we can express the various measures in a different manner revealing their hierarchy.

2.4.1 Hierarchy of Distance Measures Revisited

Let $(R, T) \rightarrow Y^2 \in \mathrm{Img}(\Omega)^2 \cap AC(\Omega; \mu^2)$ be a pair of absolutely μ^2-continuous images. As in the proof Theorem 2.19 we can apply the general transformation rule (cf. Theorem A.5 in the Appendix). Therefore, we can rewrite the integrals from the definition of the SSD and CC as μ^2-integrals over Y^2. We obtain the following representations:

$$\tfrac{2}{|\Omega|} \cdot \mathcal{D}^{\mathrm{SSD}}[R, T] = \int_{Y \times Y} p_{RT}(r, t)\, (r - t)^2 \; \mathrm{d}\mu^2(r, t)$$

$$-\tfrac{\mathrm{sign}(\mathcal{D}^{\mathrm{CC}}[R,T])}{|\Omega|} \cdot \mathcal{D}^{\mathrm{CC}}[R, T] = \int_{Y \times Y} p_{RT}(r, t)\, \frac{(r - \mu[R])(t - \mu[T])}{\sigma[R]\,\sigma[T]} \; \mathrm{d}\mu^2(r, t)$$

$$-\mathcal{D}^{\mathrm{MI}}[R, T] = \int_{Y \times Y} p_{RT}(r, t)\, \log \frac{p_{RT}(r, t)}{p_R(r)\, p_T(t)} \; \mathrm{d}\mu^2(r, t)$$

Basically, the distance measures are weighted means of distances of gray values.

The SSD measures the distance by the difference of intensities $(r - t)^2$. Note that the images do not take part in this difference. The information about the images is brought into the functional only by the weighting with the joint density p_{RT}.

The cross-correlation is similar to the SSD, but normalizes the gray values w.r.t. mean and variance. Therefore, information from images are brought into the distance term. Anyhow, $\mu[R], \mu[T], \sigma[R]$, and $\sigma[T]$ are constants that affect all gray values r, t in the same way. Thus, the CC is not substantially more general than the SSD.

Finally, the MI compares in place of intensities statistics based on the images. This allows on one hand, for a structural comparison of nearly any images and modalities, respectively. On the other hand, the generality of the MI distance measure is hard to control.

So far we considered the general case of absolutely μ-continuous images with μ an arbitrary measure. Nevertheless, for the design of a numerical method based on MI we have to choose a particular measure for μ. Two natural choices are Lebesgue and counting measure.

The numerical methods presented in the latter chapters are based on derivatives of both the images and densities. Therefore, images and densities must be continuous and continuously differentiable, respectively. The range of any non-constant continuous image is a real interval and hence uncountable. Thus, for these images the counting measure does not apply in general and the Lebesgue measure is more suitable for us. To this end, next we consider absolutely continuous images. Subsequently, we also treat absolutely #-continuous images. Even if they are not suitable for us, mutual information w.r.t. the counting measure is frequently used in literature [9, 10, 14, 39, 45, 60].

2.5 MI of Absolutely Continuous Images

In this section we consider absolutely continuous images, though images
that possess a density w.r.t. the Lebesgue measure. The MI for an ab-
solutely continuous pair of images $(R, T) \in \mathrm{Img}(\Omega)^2 \cap AC(\Omega; \mathcal{L}^2)$ is given
by

$$\mathrm{MI}[R, T] = \int_{\mathbb{R} \times \mathbb{R}} p_{RT}(r, t) \log \frac{p_{RT}(r, t)}{p_R(r)\, p_T(t)} \, \mathrm{d}(r, t).$$

The crucial points are the existence and computation of densities. Next
we give necessary conditions for the existence of densities in order to
characterize absolutely continuous images. The remaining part is devoted to their computation.

2.5.1 Existence of Densities

If a measure possesses a density, then it can be computed by an integral
w.r.t. another measure. Here, if the joint distribution P_{RT} of two images
R, T is absolutely continuous, then we can compute $P_{RT}(A)$ by a Lebesgue
integral, namely

$$P_{RT}(A) = \int_A p_{RT}(r, t) \, \mathrm{d}(r, t).$$

Now let us assume $A = \{(a)\}$, $a \in \mathbb{R}^2$ is a singleton and $P_{RT}(\{a\}) > 0$. Then
P_{RT} cannot be computed by a Lebesgue integral, since $\{a\}$ is a *null set*,
i.e., a set with Lebesgue measure zero and therefore

$$\int_{\{a\}} g(r, t) \, \mathrm{d}(r, t) = 0 \quad \text{for all } g \in L^1(\mathbb{R}^2).$$

Hence, P_{RT} possesses no density. Consequently, a basic necessity for the
existence of a density is that whenever $P_{RT}(A) > 0$ then $\mathcal{L}^2(A) > 0$.

Note, this implies that images with a discrete range are not absolutely
continuous. For a pair of images with discrete range the joint distribution
takes positive values for countable sets that are alway null sets w.r.t. the
Lebesgue measure.

It turns out, the above condition is not only a necessity but still equivalent
to the existence of a density.

Theorem 2.20

Let $R, T \in \mathrm{Img}(\Omega)$ be images with joint distribution P_{RT}. Then P_{RT} is
absolutely continuous if and only if

$$\mathcal{L}^2(A) = 0 \quad \Rightarrow \quad P_{RT}(A) = 0 \quad \text{for all } A \in \mathcal{B}(\mathbb{R}^2).$$

Proof.   The Lebesgue measure is σ-finite. Therefore, we can apply the theorem by Radon-Nikodym (cf. Theorem A.6 on page 160) that yields the theorem.   ∎

The above theorem characterizes absolutely continuous distributions. Nevertheless, we are interested in the images and pairs of images respectively for which the joint density exists. Unfortunately there is no general characterization as for the absolute continuity of distributions. So we give some necessary conditions for the existence of a joint density derived from previous results.

Lemma 2.21 (Necessary Conditions for Absolute Continuity)
Let $\Omega \subset \mathbb{R}^d$ and $(R,T) \in \mathrm{Img}(\Omega)^2 \cap AC(\Omega; \mathcal{L}^2)$ be absolutely continuous images. Then the following statements hold:

a) $R,T \in AC(\Omega; \mathcal{L})$ (existence of marginals)

b) If there exists a constant $c \in \mathbb{R}$ and a set $S \subseteq \Omega$ such that $R(x) = c$ or $T(x) = c$ for all $x \in S$, then $\mathcal{L}^d(S) = 0$.   (non-constant)

c) Let $\xi : \mathbb{R} \to \mathbb{R}$ and $\Gamma_\xi := \{(y, \xi(y)) : y \in \mathbb{R}\}$ be its graph with $\mathcal{L}^2(\Gamma_\xi) = 0$. If $\xi(R(x)) = T(x)$ or conversely $R(x) = \xi(T(x))$ holds for all x in a set $S \subseteq \Omega$, then $\mathcal{L}^d(S) = 0$.   (independence)

Proof.   a) Direct consequence of Lemma 2.12.
b) Since $0 = \mathcal{L}(\{c\}) = \mathcal{L}^2(\{c\} \times \mathbb{R})$ Theorem 2.20 and Lemma 2.12 imply $0 = P_{RT}(\{c\} \times \mathbb{R}) = P_R(\{c\}) = P(R^{-1}(\{c\})) \geq P(S)$ and therefore $\mathcal{L}^d(S) = 0$. The proof for $T(x) = c$ on S is along the same line.
c) Let $\xi(R(x)) = T(x)$ on S. As in the proof of part b) from $\mathcal{L}^2(\Gamma_\xi) = 0$ follows

$$
\begin{aligned}
0 = P_{RT}(\Gamma_\xi) &= P((R,T)^{-1}(\Gamma_\xi)) \\
&= P(\{x \in \Omega : \exists (y, \xi(y)) \in \Gamma_\xi : R(x) = y \wedge T(x) = \xi(y)\}) \\
&\geq P(\{x \in S : \exists (y, \xi(y)) \in \Gamma_\xi : R(x) = y \wedge T(x) = \xi(y)\}) \\
&= P(\{x \in S : \exists (y, \xi(y)) \in \Gamma_\xi : R(x) = y \wedge \xi(R(x)) = \xi(y)\}) \\
&= P(\{x \in S : \exists y \in \mathbb{R} : R(x) = y\}) \\
&= P(S).
\end{aligned}
$$

Hence $P(S) = 0$ and therefore $\mathcal{L}^d(S) = 0$. The proof for $R(x) = \xi(T(x))$ follows in an analogous manner.   ∎

The above Lemma gives some quite restrictive characterizations for the class of absolutely continuous images. These include:

- Images with constant regions possess no densities w.r.t. the Lebesgue measure.

- Let $R \in \mathrm{Img}(\Omega)$. Then $(R, R) \notin AC(\Omega; \mathcal{L}^2)$ and hence $\mathrm{MI}[R, R]$ is undefined.

- As motivated in the beginning, two images $R, T \in \mathrm{Img}(\Omega)$ are considered to be most similar if a function $\xi : \mathbb{R} \to \mathbb{R}$ exists such that $\xi(R) = T$ or vice versa. But then $(R, T) \notin AC(\Omega; \mathcal{L}^2)$ and therefore $\mathrm{MI}[R, T]$ is undefined.

This is somehow a negative result since many images of practical interest can neither be measured by the MI w.r.t. the Lebesgue integral nor by the MI w.r.t. the counting measure. However, there is a way to enforce absolute continuity of arbitrary images. Therefore, we simply add a little "noise".

Theorem 2.22

Let $R, T \in \mathrm{Img}(\Omega)$ *be two images and* $(\eta_1, \eta_2) \in AC(\Omega; \mathcal{L}^2)$ *with density* $K \in L^1(\mathbb{R}^2)$. *If* (R, T) *and* (η_1, η_2) *are stochastically independent, then* $(R + \sigma\eta_1, T + \sigma\eta_2) \in AC(\Omega; \mathcal{L}^2)$ *for all* $\sigma > 0$ *with density*

$$p(r, t) = \frac{1}{|\Omega|} \int_\Omega \frac{1}{\sigma^2} K\left(\frac{r - R(x)}{\sigma}, \frac{t - T(x)}{\sigma}\right) \, \mathrm{d}x.$$

Proof. To simplify presentation let $\eta_\sigma := \sigma(\eta_1, \eta_2)$ with distribution P_{η_σ} and density p_{η_σ}. The scaling of η_1, η_2 with $\sigma > 0$ does not change the stochastic independence from (R, T), such that η_σ and (R, T) are stochastically independent, too.

Furthermore $\eta_\sigma^{-1}(A) = (\sigma\eta_1, \sigma\eta_2)^{-1}(A) = (\eta_1, \eta_2)^{-1}(\sigma^{-1}A)$, such that

$$P_{\eta_\sigma}(A) = P_{\eta_1\eta_2}(\sigma^{-1}A)) = \int_{\sigma^{-1}A} K(r, t) \, \mathrm{d}(r, t) = \int_A \frac{1}{\sigma^2} K(r/\sigma, t/\sigma) \, \mathrm{d}(r, t).$$

Hence P_{η_σ} is absolutely continuous with density $p_{\eta_\sigma}(r, t) := \frac{1}{\sigma^2} K(r/\sigma, t/\sigma)$ and therefore $\eta_\sigma \in AC(\Omega; \mathcal{L}^2)$. For (R, T) and stochastically independent absolutely continuous η_σ it can be shown that the sum $(R, T) + \eta_\sigma$ is absolutely continuous with density

$$p(r, t) = \int_{\mathbb{R}^2} p_{\eta_\sigma}(r - \varrho, t - \vartheta) \, \mathrm{d}P_{RT}(\varrho, \vartheta).$$

For a proof see [31, §4.6]. Finally, applying the general transformation rule given in Theorem A.5 yields

$$p(r, t) = \int_\Omega p_{\eta_\sigma}(r - R(x), t - T(x)) \, \mathrm{d}P(x) = \frac{1}{|\Omega|} \int_\Omega p_{\eta_\sigma}(r - R(x), t - T(x)) \, \mathrm{d}x.$$

∎

The above theorem is important. First, it allows for the measurement of MI for biased versions of arbitrary images. Second, it gives an explicit formula for the computation of the density. The numerical methods in the latter chapters are based on this result. It provides a quite practical way to overcome theoretical problems and allows to ensure further wanted properties on the density, e.g., differentiability, as we will show in chapter 3.

Nevertheless, densities are no magic objects. In the following section we give explicit formulae for densities of functions that are smooth and regular in some sense.

2.5.2 Computing Densities of Smooth Functions

In general, the density of an absolutely continuous function can be computed by differentiating the so-called *distribution function*. The distribution function $F : \mathbb{R}^k \to [0,1]$ of a function $f \in L^1(\Omega)^k$ is defined as

$$F(x_1, x_2, \ldots, x_k) := P_f(]-\infty, x_1] \times]-\infty, x_2] \times \ldots \times]-\infty, x_k]),$$

whereby P_f denotes the distribution of f. If f is absolutely continuous it can be shown that F is differentiable and the density of f is given by

$$p_f(x_1, x_2, \ldots, x_k) = \frac{\partial^k}{\partial x_1 \partial x_2 \ldots \partial x_k} F(x_1, x_2, \ldots, x_k).$$

Nevertheless, this result is of little use for practical computations. Furthermore, there is no direct link between a density and the function that induces the distribution.

In the following, we consider absolutely continuous functions that are also continuously differentiable. There are explicit formulae for the densities if these functions satisfy a regularity constraint that allows for inversion in a sense.

The main result presented in this section will be quite general and is a profound result from geometric measure theory. In particular, dealing with functions that map between spaces with different dimensions is difficult and involved. This occurs, for example, when we want to register three-dimensional images. Then, for images $R, T \in \text{Img}(\Omega)$ with $\Omega \subset \mathbb{R}^3$ we have to compute the density of the joint function $(R, T) : \Omega \to \mathbb{R}^2$ that maps from three to two dimensions. To point out the main ideas we will give a simple theorem for the equi-dimensional case with a constructive proof.

Definition 2.23 (Jacobian Matrix)

Let $\Omega \subseteq \mathbb{R}^d$ and $f : \Omega \to \mathbb{R}^k$. If f is differentiable at $x \in \Omega$ the matrix defined as

$$\mathrm{D}f(x) := \begin{pmatrix} \frac{\partial f_1}{\partial x_1}(x) & \frac{\partial f_1}{\partial x_2}(x) & \cdots & \frac{\partial f_1}{\partial x_d}(x) \\ \frac{\partial f_2}{\partial x_1}(x) & \frac{\partial f_2}{\partial x_2}(x) & \cdots & \frac{\partial f_2}{\partial x_d}(x) \\ \vdots & \vdots & & \vdots \\ \frac{\partial f_k}{\partial x_1}(x) & \frac{\partial f_k}{\partial x_2}(x) & \cdots & \frac{\partial f_k}{\partial x_d}(x) \end{pmatrix} \in \mathbb{R}^{k \times d}$$

is called Jacobian matrix of f at x.

The problem of finding densities if the underlying functions are differentiable is strongly related to the change variables in integration. We will illustrate this relation in the following example.

Example 2.24

Let $\Omega, Y \subset \mathbb{R}^d$ be open sets and $f : \Omega \to Y$ continuously differentiable. If f is one-to-one then we can apply the well-known transformation rule for the Lebesgue integral (cf. Forster 3) for $g \in L^1(Y)$. We recall that the transformation rule reads as

$$\int_Y g(y)\, \mathrm{d}y = \int_\Omega |\det \mathrm{D}f(x)|\, g(f(x))\, \mathrm{d}x.$$

For $g \circ f \in L^1(\Omega)$ we apply the transformation rule with f^{-1}. Thus

$$\int_\Omega g(f(x))\, \mathrm{d}x = \int_Y |\det \mathrm{D}f^{-1}(y)|\, g(f(f^{-1}(y)))\, \mathrm{d}y$$
$$= \int_Y \frac{1}{|\det \mathrm{D}f(f^{-1}(y))|}\, g(y)\, \mathrm{d}y$$

whereby we use the identity $\det \mathrm{D}f^{-1}(y) = 1/\det \mathrm{D}f(f^{-1}(y))$ following from $1 = |\det \mathrm{D}(f(f^{-1}(y)))| = |\det \mathrm{D}f(f^{-1}(y))| \cdot |\det \mathrm{D}f^{-1}(y)|$. Hence

$$\int_Y g(y)\, \mathrm{d}P_f(x) = \frac{1}{|\Omega|} \int_\Omega g(f(x))\, \mathrm{d}x = \int_Y p_f(y)\, g(y)\, \mathrm{d}y \qquad (2.6)$$

and so P_f is absolutely continuous with density

$$p_f(y) = \begin{cases} \left. \frac{1/|\Omega|}{|\det \mathrm{D}f(x)|} \right|_{x = f^{-1}(y)} & \text{if } y \in Y \\ 0 & \text{if } y \notin Y. \end{cases} \qquad (2.7)$$

In the above example we computed the density of an invertible function. Next, we generalize this result for functions that are piecewise one-to-one. The idea is quite simple. We separate the domain into pieces where the function is one-to-one, compute the densities for each piece as shown above, and finally collect them.

Theorem 2.25

Let $\Omega, Y \subset \mathbb{R}^d$ and $f : \Omega \to Y$ be continuously differentiable. If there exists a family $\{W_n\}_{n \in \mathbb{N}}$ of pairwise disjoint open sets such that $N := \Omega \setminus (\cup_{n \in \mathbb{N}} W_n)$ is a null set, that is $\mathcal{L}^d(N) = 0$ and the restrictions $f_n := f|_{W_n}$ are one-to-one, then P_f is absolutely continuous and its density is given by

$$p_f(y) = \begin{cases} \displaystyle\sum_{x \in f^{-1}(\{y\}) \setminus N} \dfrac{1/|\Omega|}{|\det \mathrm{D}f(x)|} & \text{if } y \in Y \\ 0 & \text{if } y \notin Y. \end{cases}$$

Proof. Since $\{W_n\}_{n \in \mathbb{N}} \cup \{N\}$ is a partition of Ω, we have for arbitrary $g : Y \to \mathbb{R}$ with $g \circ f \in L^1(\Omega)$

$$\int_\Omega g(f(x))\, \mathrm{d}x = \sum_{n \in \mathbb{N}} \int_{W_n} g(f(x))\, \mathrm{d}x = \sum_{n \in \mathbb{N}} \int_{W_n} g(f_n(x))\, \mathrm{d}x.$$

Let $Y_n := f(W_n)$. Then, as shown in the above example, from (2.6) and (2.7) follows $\frac{1}{|\Omega|} \int_{W_n} g(f_n(x))\, \mathrm{d}x = \int_{Y_n} p_{f_n}(y)\, g(y)\, \mathrm{d}y$ with

$$p_{f_n}(y) = \begin{cases} \left. \dfrac{1/|\Omega|}{|\det \mathrm{D}f(x)|} \right|_{x = f_n^{-1}(y)} & \text{if } y \in Y_n \\ 0 & \text{if } y \notin Y_n. \end{cases}$$

Thus

$$\sum_{n \in \mathbb{N}} \int_{W_n} g(f_n(x))\, \mathrm{d}x = \sum_{n \in \mathbb{N}} \int_{Y_n} p_{f_n}(y)\, g(y)\, \mathrm{d}y = \int_Y \left(\sum_{n \in \mathbb{N}} p_{f_n}(y) \right) g(y)\, \mathrm{d}y$$

and

$$\sum_{n \in \mathbb{N}} p_{f_k}(y) = \sum_{n \in \mathbb{N} : y \in Y_n} p_{f_n}(y) = \sum_{x \in \Omega : f_n^{-1}(y) = x \,\wedge\, y \in Y_n} \dfrac{1/|\Omega|}{|\det \mathrm{D}f_n(x)|}.$$

Finally

$$\begin{aligned} \{x \in \Omega : \exists n : f_n^{-1}(y) = x \,\wedge\, y \in Y_n\} &= \{x \in \Omega : f(x) = y \,\wedge\, \exists n : y \in Y_n\} \\ &= \{x \in \Omega : f(x) = y\} \setminus N \\ &= f^{-1}(\{y\}) \setminus N. \end{aligned}$$

■

The above theorem gives a formula for the densities of functions that map between equi-dimensional spaces. Hence, if we want to apply the theorem for the computation of the joint density of two images, the images must be two-dimensional.

For a generalization of Theorem 2.25 to functions $f : \Omega \subseteq \mathbb{R}^d \to \mathbb{R}^k$ with $d > k$ and therefore to functions that map a higher to lower dimensional space we have to solve two major problems. First, the Jacobian matrix is no longer a square matrix and therefore its determinant is not defined. Second, the preimage of a function value is in general an uncountable set. Therefore, we cannot use an ordinary sum.

The first problem is solved by the following generalization of the Jacobian.

Definition 2.26 (Absolute Jacobian)

Let $\Omega \subseteq \mathbb{R}^d$ and $f : \Omega \to \mathbb{R}^k$ with $k \leq d$. The absolute Jacobian J_f of f at x is defined as

$$J_f(x) := \begin{cases} \sqrt{\det(\mathrm{D}f(x)\,\mathrm{D}f(x)^\top)} & \text{if } f \text{ is differentiable at } x \\ \infty & \text{if } f \text{ is not differentiable at } x \end{cases}$$

Note that the above definition includes the equi-dimensional case. If $k = d$ in the above definition, the Jacobian matrix is a square matrix yielding

$$\sqrt{\det(\mathrm{D}f(x)\,\mathrm{D}f(x)^\top)} = \sqrt{\det \mathrm{D}f(x) \cdot \det \mathrm{D}f(x)^\top}$$
$$= \sqrt{(\det \mathrm{D}f(x))^2}$$
$$= |\det \mathrm{D}f(x)|.$$

Furthermore, for images $R, T \in \mathrm{Img}(\Omega)$ we have

$$J_{(R,T)} = \sqrt{||\nabla R||^2\,||\nabla T||^2 - \langle \nabla R, \nabla T \rangle^2}$$
$$= ||\nabla R||\,||\nabla T||\sqrt{1 - \cos^2 \angle(\nabla R, \nabla T)}$$
$$= ||\nabla R||\,||\nabla T||\,\sin \angle(\nabla R, \nabla T).$$

Note that for three-dimensional images, this is the length of the cross-product of the gradients, i.e., $J_{(R,T)} = ||\nabla R \times \nabla T||$, where the cross-product is defined for vectors $v, w \in \mathbb{R}^3$ by

$$v \times w = (v_2 w_3 - v_3 w_2, v_3 w_1 - v_1 w_3, v_1 w_2 - v_2 w_1)^\top.$$

The second problem was that the preimage is no longer a countable set that can be measured by a sum. The generalization of this is much more involved. Therefore, we have to introduce the m-dimensional Hausdorff measure.

Definition 2.27 (Hausdorff Measure)

Let $m \in \mathbb{N}$. Then   the m-dimensional volume of a set $A \subseteq \mathbb{R}^d$ is defined as

$$\text{vol}^m(A) := \frac{\pi^{m/2} \left(\frac{1}{2} \, \text{diam}(A)\right)^m}{\Gamma(1 + \frac{m}{2})}$$

with $\text{diam}(A) := \sup\{\|x - y\|_{\mathbb{R}^d} : x, y \in A\}$ *for* $A \neq \emptyset$, $\text{diam}(\emptyset) := 0$ *and the so-called Gamma function* $\Gamma(x) := \int_0^\infty t^{x-1} e^{-t} \, \mathrm{d}t$. *Defining*

$$\mathcal{H}_\varepsilon^m(A) := \inf\left\{\sum_{i=1}^\infty \text{vol}^m(B_i) \; : \; B_i \subseteq \mathbb{R}^d, \text{diam}(B_i) \leq \varepsilon \text{ and } A \subseteq \bigcup_{i=1}^\infty B_i\right\}.$$

the m-dimensional Hausdorff measure $\mathcal{H}^m$ is given by

$$\mathcal{H}^m(A) := \lim_{\varepsilon \to 0} \mathcal{H}_\varepsilon^m(A).$$

The m-dimensional Hausdorff measure is also called *surface measure* and therefore still denoted by S. In the following chapter we will use the second notation for boundary integrals $\int_{\partial\Omega} f(x) \, \mathrm{d}S(x)$ over the boundary $\partial\Omega$ of a domain $\Omega \subset \mathbb{R}^d$. In terms of the notation used here a boundary integral reads

$$\int_{\partial\Omega} f(x) \, \mathrm{d}S(x) = \int_{\partial\Omega} f(x) \, \mathrm{d}\mathcal{H}^{d-1}(x).$$

The Hausdorff measure enables us to measure lower dimensional subsets of $\mathbb{R}^d$. In particular, the zero-dimensional Hausdorff measure $\mathcal{H}^0$ is the counting measure $\#$, the one-dimensional Hausdorff measure $\mathcal{H}^1$ is length, and the d-dimensional Hausdorff measure $\mathcal{H}^d$ is the d-Lebesgue measure $\mathcal{L}^d$.

Now we are able to formulate a more general version of the above theorem.

Theorem 2.28

Let $\Omega \subseteq \mathbb{R}^d$, $f : \Omega \to \mathbb{R}^k$ be a measurable function with $k \leq d$, and the set of singular points $\Sigma_f := \{x \in \Omega : J_f(x) \in \{0, \infty\}\}$. If

$$\mathcal{L}^d(\Sigma_f) = 0 \quad \text{and} \quad \mathcal{L}^k(\{y \in f(\Sigma_f) : \mathcal{H}^{d-k}(f^{-1}(\{y\}) \cap \Sigma_f) > 0\}) = 0$$

then f is absolutely continuous with density

$$p_f(y) = \begin{cases} \displaystyle\int_{f^{-1}(y)} \frac{1/|\Omega|}{J_f(x)} \, \mathrm{d}\mathcal{H}^{d-k}(x) & \text{if } y \in f(\Omega), \\ 0 & \text{if } y \notin f(\Omega). \end{cases}$$

Proof.   For a proof see [31, §8] and references therein. ∎

In view of Theorem 2.25 for the equi-dimensional case $k = d$ the require-ment $\mathcal{L}^k(\{y \in f(\Sigma_f) : \mathcal{H}^{d-k}(f^{-1}(\{y\}) \cap \Sigma_f) > 0\}) = 0$ of the theorem comes to $\mathcal{L}^d(\Sigma_f) = 0$ since $\mathcal{H}^{d-k} = \mathcal{H}^0 = \#$ such that $\mathcal{H}^{d-k}(f^{-1}(\{y\}) \cap \Sigma_f) > 0$ holds for all $y \in \Sigma_f$. Furthermore, the set N in Theorem 2.25 is a superset of Σ_f such that $\mathcal{L}^d(N) = 0$ implies $\mathcal{L}^d(\Sigma_f) = 0$.

In particular, for two and three-dimensional images we have the following corollary.

Corollary 2.29

Let $\Omega \subset \mathbb{R}^d$ and $(R, T) \in \text{Img}(\Omega)^2$ be an image pair. If (R, T) fulfill the assumptions of Theorem 2.28 then $(R, T) \in AC(\Omega; \mathcal{L}^2)$ and for two-dimensional images $(d = 2)$ the joint density is given by

$$p_{RT}(r, t) = \begin{cases} \dfrac{1}{|\Omega|} \displaystyle\sum_{N(r,t)} \dfrac{1}{J(x)} & \text{if } (r, t) \in R(\Omega) \times T(\Omega), \\ 0 & \text{else,} \end{cases}$$

for three-dimensional images $(d = 3)$ the joint density reads

$$p_{RT}(r, t) = \begin{cases} \dfrac{1}{|\Omega|} \displaystyle\int_{N(r,t)} \dfrac{1}{J(x)} \, dS(x) & \text{if } (r, t) \in R(\Omega) \times T(\Omega), \\ 0 & \text{else.} \end{cases}$$

with level sets $N(r, t) := \{x \in \Omega : R(x) = r \wedge T(x) = t\}$ and absolute Jacobian $J := \sqrt{\|\nabla R\|^2 \|\nabla T\|^2 - \langle \nabla R, \nabla T \rangle^2}$.

2.6 Mutual Information of Discrete Images

In this section we will analyze the case that images take only a finite or at most countable number of gray values. We will show that always a (joint) density exists with an explicit representation. Furthermore, the μ-integrals become sums. These simplifications ease up the analysis of MI and many difficulties as in the general or non-discrete case do not occur.

Definition 2.30 (Discrete Image)

A function $f : \Omega \to Y$ is called discrete if Y is countable.

Discrete images take only a countable number of gray values. Note that it is not required that the domain Ω is discrete, too. An example for discrete functions are step functions.

> **Lemma 2.31**
>
> Let $f : \Omega \to Y$ be a measurable function with distribution P_f. Then f is discrete if and only if $f \in AC(\Omega; \#)$ with $\#$-density p_f is given by $p_f(y) = P_f(\{y\})$, $y \in Y$.

Proof. "$\Rightarrow$". Let $A \subset Y$. Since, Y is countable A is countable as well and therefore

$$P_f(A) = \sum_{a \in A} P_f(\{a\}) = \int_A P_f(\{a\}) \, \mathrm{d}\#(a) = \int_A p_f(a) \, \mathrm{d}\#(a).$$

Thus, $f \in AC(\Omega; \#)$ with $\#$-density $p_f = P_f(\{\cdot\})$.

"$\Leftarrow$". For $f : \Omega \to Y \in AC(\Omega; \#)$ we have

$$P_f(Y) = \int_Y p_f(y) \, \mathrm{d}\#(y) = \sum_{y \in Y} p_f(y).$$

Hence, Y must be countable and therefore f discrete. ∎

The lemma implies that the space of discrete images is given by $\mathrm{Img}(\Omega) \cap AC(\Omega; \#)$. More important, it implies the density of discrete images always exists.

For the computation of the MI we are interested in the joint density of a pair of images. The existence of a joint density implies the existence of its marginals (cf. Lemma 2.12), conversely this does not hold in general. For example, let R be an image that is absolutely continuous and therefore possesses a density w.r.t. the Lebesgue measure. Then we saw in the previous section the joint density for (R, R) does not exist (cf. Lemma 2.21). However, the joint density of two discrete images always exists.

> **Corollary 2.32**
>
> Let $R, T \in \mathrm{Img}(\Omega) \cap AC(\Omega, \#)$ be two discrete images. Then (R, T) is discrete and $(R, T) \in \mathrm{Img}(\Omega)^2 \cap AC(\Omega; \#^2)$.

Proof. Without loss of generality let $Y \subset \mathbb{R}$ be both the range of R and T. Since R, T are discrete images Y must be countable by definition. Hence, $Y \times Y$ is countable, too, and therefore (R, T) is discrete. Since $\mathrm{Img}(\Omega) \subset L^1(\Omega)$, the images R, T and (R, T) are integrable and therefore measurable. Applying Lemma 2.31 yields the corollary. ∎

The MI of discrete images $R, T : \Omega \to Y$ is given by

$$\mathrm{MI}[R, T] = \sum_{r \in Y} \sum_{t \in Y} P_{RT}(\{(r, t)\}) \log \frac{P_{RT}(\{(r, t)\})}{P_R(\{r\}) \, P_T(\{t\})}.$$

Note that the general results derived in the previous sections hold, since considering discrete images are the particular case for choosing the counting measure as the "reference" measure μ. In particular, the representation (2.19) remains an integral over Ω, given by

$$\mathrm{MI}[R,T] = \frac{1}{|\Omega|} \int_\Omega \log \frac{P_{RT}(\{R(x),T(x)\})}{P_R(\{R(x)\})\,P_T(\{T(x)\})} \, \mathrm{d}x.$$

Additionally, for the particular case of discrete images the MI is bounded from above by the entropy of the images. For discrete images the entropies $H[R]$ and $H[T]$ reads

$$H[R] = -\sum_r P_R(\{r\}) \log P_R(\{r\}) \quad \text{and} \quad H[T] = -\sum_t P_T(\{t\}) \log P_T(\{t\}).$$

Theorem 2.33

Let $R,T \in \mathrm{Img}(\Omega) \cap AC(\Omega; \#)$ be discrete images. Then

$$\mathrm{MI}[R,T] \leq \min\{H[R], H[T]\}$$

with equality if a one-to-one mapping $\xi : \mathbb{R} \to \mathbb{R}$ exists with $T = \xi(R)$ or $R = \xi(T)$, respectively.

Proof. We only show $\mathrm{MI}[R,T] \leq H[R]$. The proof for $\mathrm{MI}[R,T] \leq H[T]$ is along the same line. First, we observe $P_T(t) = \sum_r P_{RT}(\{(r,t)\})$. Since P_{RT} is non-negative we have $P_T(t) \geq P_{RT}((r,t))$ for all r,t. Thus

$$\log \frac{P_{RT}}{P_R P_T} \leq \log \frac{P_T}{P_R P_T} = \log \frac{1}{P_R} = -\log P_R$$

and therefore

$$\mathrm{MI}[R,T] = \sum_{r,t} P_{RT} \log \frac{P_{RT}}{P_R P_T} \leq -\sum_{r,t} P_{RT} \log P_R = -\sum_r P_R \log P_R = H[R].$$

Furthermore, if $T = \xi(R)$ with one-to-one $\xi : \mathbb{R} \to \mathbb{R}$ then

$$\begin{aligned}
P_{RT}(\{(r,t)\}) &= P((R,T)^{-1}(\{(r,t)\})) \\
&= P(\{x : R(x) = r \wedge T(x) = t\}) \\
&= P(\{x : \xi(T(x)) = r \wedge T(x) = t\}) \\
&= P(\{x : T(x) = \xi^{-1}(r) \wedge T(x) = t\})
\end{aligned}$$

and therefore

$$P_{RT}(\{(r,t)\}) = \begin{cases} P_T(\{t\}) & \text{if } t = \xi^{-1}(r), \\ 0 & \text{else.} \end{cases}$$

Thus, the joint entropy becomes the entropy of T, i.e.,

$$H[(R,T)] = -\sum_{r,t} P_{RT} \log P_{RT} = -\sum_t P_T \log P_T = H[T]$$

and $\mathrm{MI}[R,T] = H[R] + H[T] - H[(R,T)] = H[R] + H[T] - H[T] = H[R]$.   ∎

Note that we consider two images R and T to be most similar if a mapping ξ exists with $T = \xi(R)$ or $R = \xi(T)$. Thus, the theorem states at least for ξ one-to-one the MI is maximal.

2.6.1 Relations to Absolutely Continuous Images

Finally, we investigate the relation between discrete and absolutely continuous images. Therefore, we consider discretized versions of continuous images.

Let $I_{\beta,k} := (k\beta - \frac{\beta}{2}, k\beta + \frac{\beta}{2}]$, $k \in \mathbb{Z}$ be intervals with fixed width $\beta > 0$. We define the discrete versions $R_\beta, T_\beta : \Omega \to \mathbb{Z}$ of images $R, T \in \mathrm{Img}(\Omega)$ as

$$R_\beta(x) := k \text{ if } R(x) \in I_{\beta,k} \quad \text{and} \quad T_\beta(x) := \ell \text{ if } T(x) \in I_{\beta,\ell}.$$

Then R_β, T_β are absolutely continuous w.r.t. the counting measure with densities

$$p_{R_\beta}(k) = P_{R_\beta}(\{k\}) = P_R(I_{\beta,k}), \quad p_{T_\beta}(\ell) = P_{T_\beta}(\{\ell\}) = P_T(I_{\beta,\ell}),$$

and joint density

$$p_{R_\beta T_\beta}(k,\ell) = P_{R_\beta T_\beta}(\{(k,\ell)\}) = P_{RT}(I_{\beta,k} \times I_{\beta,\ell}).$$

Theorem 2.34

Let $(R,T) \in \mathrm{Img}(\Omega)^2 \cap AC(\Omega; \mathcal{L}^2)$ be absolutely continuous with joint density p_{RT}. If p_{RT} is continuous and Riemann integrable, then

$$\lim_{\beta \to 0} \mathrm{MI}[R_\beta, T_\beta] = \mathrm{MI}[R,T].$$

Proof.   For continuous $p_{R_\beta T_\beta}$ the mean value theorem ensures the existence of points $x_k \in I_{\beta,k}$, $y_\ell \in I_{\beta,\ell}$, and $(x'_k, y'_\ell) \in I_{\beta,k} \times I_{\beta,\ell}$ with

$$p_{R_\beta}(k) = \int_{\beta k - \frac{\beta}{2}}^{\beta k + \frac{\beta}{2}} p_R(x)\,\mathrm{d}x = \beta\, p_R(x_k), \quad p_{T_\beta}(\ell) = \int_{\beta k - \frac{\beta}{2}}^{\beta k + \frac{\beta}{2}} p_T(y)\,\mathrm{d}y = \beta\, p_T(x_\ell),$$

and

$$p_{R_\beta T_\beta}(k,\ell) = \int_{\beta k - \frac{\beta}{2}}^{\beta k + \frac{\beta}{2}} \int_{\beta \ell - \frac{\beta}{2}}^{\beta \ell + \frac{\beta}{2}} p_{RT}(x,y)\,\mathrm{d}x\mathrm{d}y = \beta^2\, p_{RT}(x'_k, y'_\ell).$$

Then the Riemann integratability of p_{RT} yields

$$\lim_{\beta \to 0} \mathrm{MI}[R_\beta, T_\beta] = \lim_{\beta \to 0} \sum_{k,\ell \in \mathbb{Z}} \beta^2\, p_{RT}(x'_k, y'_\ell) \log \frac{p_{RT}(x'_k, y'_\ell)}{p_R(x_k) p_T(y_\ell)}$$

$$= \int_{\mathbb{R}^2} p_{RT}(x,y) \log \frac{p_{RT}(x,y)}{p_R(x) p_T(y)}\,\mathrm{d}x\mathrm{d}y$$

$$= \mathrm{MI}[R,T].$$

          ∎

The above theorem states that the mutual information of the discretized images is consistent with the MI of the continuous images. This seems quite natural, since $\beta(R_\beta, T_\beta) \to (R, T)$ for $\beta \to 0$. But as to be expected, not every functional behaves consistent for discrete and continuous images. An example for that is the entropy of images given in Definition 2.16 on page 26.

Theorem 2.35

Let $R \in \text{Img}(\Omega \cap AC(\Omega; \mathcal{L}))$ be absolutely continuous with density p_R. If p_R is continuous and Riemann integrable, then

$$\lim_{\beta \to 0} (H[R_\beta] + \log \beta) = H[R].$$

Proof.   As in the proof above the mean value theorem ensures the existence of points $x_k \in I_{\beta,k}$, such that $p_{R_\beta}(k) = P_R(I_{\beta,k}) = \beta \, p_R(x_k)$. Thus

$$H[R_\beta] = -\sum_{k \in \mathbb{Z}} \beta \, p_R(x_k) \log(\beta \, p_R(x_k)) = -\sum_{k \in \mathbb{Z}} \beta \, p_R(x_k) \log p_R(x_k) - \log \beta$$

and hence $H[R_\beta] + \log \beta = -\displaystyle\sum_{k \in \mathbb{Z}} \beta \, p_R(x_k) \log p_R(x_k) \to H[R]$ for $\beta \to 0$.   ∎

Chapter 3

Practical Treatment of MI

The crucial point for image registration based on mutual information is the computation of densities. In this chapter we show how to compute approximations to the densities and mutual information, respectively, that are based on so-called *kernel functions*. Using kernel functions gives explicit formulae for the approximations as well as allows for a unified treatment of arbitrary images. Additionally, we are able to ensure certain properties to the approximations, as, e.g., differentiability.

3.1 Approximating Densities

Theorem 2.22 shows that adding a stochastically independent absolutely continuous function to an image pair results an absolutely continuous function again, even if images are not absolutely continuous. Thus arbitrary small perturbations of images ensure theoretical requirements and give a practical way out of the dilemma on the existence of densities. In the following, we will use densities of the kind specified by the next definition.

Definition 3.1 (Kernel, Approximating Densities p_{RT}^σ)

We call a function $K \in L^1(\mathbb{R}^k)$ kernel if K is non-negative, bounded from above and $\int_\mathbb{R} K(x)\,\mathrm{d}x = 1$. Furthermore, for $\sigma > 0$ we define the scaled kernel $K_\sigma(x) := \frac{1}{\sigma} K(x/\sigma)$.

For images $R, T \in \mathrm{Img}(\Omega)$, a differentiable kernel $K \in C^1(\mathbb{R})$ with $K(0) = \max_{x \in \mathbb{R}} K(x)$ and $K_\sigma(r,t) := K_\sigma(r)K_\sigma(t)$ we define the approximative densities

$$p_{RT}^\sigma(r,t) := \frac{1}{|\Omega|} \int_\Omega K_\sigma(r - R(x), t - T(x))\,\mathrm{d}x$$

and

$$p_R^\sigma(r) := \int_\mathbb{R} p_{RT}^\sigma(r,t)\,\mathrm{d}t, \quad p_T^\sigma(t) := \int_\mathbb{R} p_{RT}^\sigma(r,t)\,\mathrm{d}r.$$

Approximating densities are again densities. The kernel takes only positive values such that $p_{RT}^\sigma, p_R^\sigma$, and p_T^σ are non-negative, too. Furthermore, a direct computation shows

$$\int_{\mathbb{R}^2} p_{RT}^\sigma(r,t) \, \mathrm{d}(r,t) = \frac{1}{|\Omega|} \int_\Omega \left(\int_{\mathbb{R}^2} K_\sigma(r - R(x), t - T(x)) \, \mathrm{d}(r,t) \right) \mathrm{d}x$$

$$= \frac{1}{|\Omega|} \int_\Omega \left(\underbrace{\int_{\mathbb{R}^2} K_\sigma(r,t) \, \mathrm{d}(r,t)}_{=1} \right) \mathrm{d}x = 1$$

and therefore $\int_{\mathbb{R}} p_R^\sigma(r) \, \mathrm{d}r = \int_{\mathbb{R}} p_T^\sigma(t) \, \mathrm{d}t = 1$. Hence, $p_{RT}^\sigma, p_R^\sigma$, and p_T^σ are non-negative functions integrating to one and therefore densities of a distribution P_{RT}^σ given by

$$P_{RT}^\sigma(A) := \int_A p_{RT}^\sigma(r,t) \, \mathrm{d}(r,t).$$

Moreover $K_\sigma(x,y) = K_\sigma(x)K_\sigma(y)$ implies

$$p_R^\sigma(r) = \frac{1}{|\Omega|} \int_\Omega K_\sigma(r - R(x)) \, \mathrm{d}x \quad \text{and} \quad p_T^\sigma(t) = \frac{1}{|\Omega|} \int_\Omega K_\sigma(t - T(x)) \, \mathrm{d}x.$$

Theorem 2.22 states that P_{RT}^σ is the distribution of $(R,T) + (\eta_1, \eta_2)$ with a function $(\eta_1, \eta_2) : \Omega \to \mathbb{R}^2$ stochastically independent from (R,T) and distributed with density K_σ. The function (η_1, η_2) does not play any further role in the following. It serves for understanding what happens to the images when switching from p_{RT} to p_{RT}^σ. In general, we can interpret (η_1, η_2) as some small noise added to the images. Hence, noising the images results in smoothing the density.

Interpreting p_{RT}^σ as an approximation to the "true" joint density p_{RT} makes only sense if p_{RT} exists. In fact, we approximate the joint distribution P_{RT} in the meaning of

$$P_{RT} \approx P_{RT}^\sigma.$$

However, if the joint density exists p_{RT}^σ is actually the convolution of p_{RT} with a kernel function K.

Lemma 3.2 (Approximation of p_{RT})
Let $R,T \in \mathrm{Img}(\Omega)$ be images with joint distribution P_{RT} and p_{RT}^σ be an approximating density. If the joint density $p_{RT} \in L^1(\mathbb{R}^2)$ of P_{RT} exists, then

a) $p_{RT}^\sigma = K_\sigma * p_{RT}.$ (density approximation)

Additionally, if p_{RT} is continuous then

b) $p_{RT}^\sigma \to p_{RT}$ as $\sigma \to 0.$ (continuity)

Proof. a)

$$p_{RT}^{\sigma}(r,t) = \frac{1}{|\Omega|} \int_{\Omega} K_{\sigma}(r - R(x), t - T(x)) \, \mathrm{d}x$$

$$= \int_{\Omega} K_{\sigma}(r - R(x), t - T(x)) \, \mathrm{d}P(x)$$

$$= \int_{\mathbb{R}^2} K_{\sigma}(r - \varrho, t - \vartheta) \, \mathrm{d}P_{RT}(\varrho, \vartheta)$$

$$= \int_{\mathbb{R}^2} K_{\sigma}(r - \varrho, t - \vartheta) \, p_{RT}(\varrho, \vartheta) \, \mathrm{d}(\varrho, \vartheta)$$

$$= K_{\sigma} * p_{RT}(r,t).$$

b) From a) and the definition of K_{σ} we have

$$p_{RT}^{\sigma}(r,t) = \int_{\mathbb{R}^2} \frac{1}{\sigma^2} K\left(\frac{r - \varrho}{\sigma}, \frac{t - \vartheta}{\sigma}\right) p_{RT}(\varrho, \vartheta) \, \mathrm{d}(\varrho, \vartheta)$$

$$= \int_{\mathbb{R}^2} K(\varrho', \vartheta') \, p_{RT}(r - \sigma\varrho', t - \sigma\vartheta') \, \mathrm{d}(\varrho', \vartheta')$$

with the substitutions $\varrho' = (r - \varrho)/\sigma$, $\vartheta' = (t - \vartheta)/\sigma$. By assumption p_{RT} is continuous and integrable on $\mathbb{R}^2$. Thus, p_{RT} must be bounded from above and a constant M exists with $\sup_{x \in \mathbb{R}^2} |p_{RT}| < M$. Then $|K(\varrho', \vartheta') \, p_{RT}(r - \sigma\varrho', t - \sigma\vartheta')| \leq M \, |K(\varrho', \vartheta')|$ and $\lim_{\sigma \to 0} K(\varrho', \vartheta') \, p_{RT}(r - \sigma\varrho', t - \sigma\vartheta') = K(\varrho', \vartheta') \, p_{RT}(r, t)$. Since $M \, |K(\varrho', \vartheta')|$ is integrable the dominated convergence implies

$$\lim_{\sigma \to 0} p_{RT}^{\sigma}(r,t) = \lim_{\sigma \to 0} \int_{\mathbb{R}^2} K(\varrho', \vartheta') \, p_{RT}(r - \sigma\varrho', t - \sigma\vartheta') \, \mathrm{d}(\varrho', \vartheta')$$

$$= p_{RT}(r,t) \underbrace{\int_{\mathbb{R}^2} K(\varrho', \vartheta') \, \mathrm{d}(\varrho', \vartheta')}_{=1}.$$

■

The requirements on the kernel used for the approximative densities are quite restrictive. In general, approximating densities works for any kernel. In particular, the requirement $\int_{\mathbb{R}} K(x) \, \mathrm{d}x = 1$ is essential. However, the differentiability of the kernel yields that the approximative density is differentiable, too. The demand for separable kernel and $K(0) = \max$ are technical conditions to simplify our analysis.

Lemma 3.3 (Differentiability and Continuity of p_{RT}^{σ})

Let p_{RT}^{σ} be as in Definition 3.1. The following statements hold:

a) p_{RT}^{σ} is continuous differentiable with

$$\partial_r p_{RT}^{\sigma}(r,t) = \frac{1}{|\Omega|} \int_{\Omega} \partial_r K_{\sigma}(r - R(x), t - T(x)) \, dx$$

and

$$\partial_t p_{RT}^{\sigma}(r,t) = \frac{1}{|\Omega|} \int_{\Omega} \partial_t K_{\sigma}(r - R(x), t - T(x)) \, dx.$$

(differentiability)

b) Let $(R_n, T_n) \subset \text{Img}(\Omega)^2$ be a sequence of image pairs. If $(R,T) \in \text{Img}(\Omega)^2$ exists with $(R_n, T_n) \rightarrow (R,T)$, that is $\|(R_n, T_n) - (R,T)\|_{L^1(\Omega)} \rightarrow 0$, then

$$p_{R_n T_n}^{\sigma} \rightarrow p_{RT}^{\sigma} \quad \text{as} \quad (R_n, T_n) \rightarrow (R,T).$$

(continuity w.r.t. images)

Proof. a) For a proof see [34, p. 280].

b) By assumption the kernel K is continuous such that $K_{\sigma}(r - R_n, t - T_n) \rightarrow K_{\sigma}(r - R, t - T)$. Furthermore, K is a kernel and therefore non-negative and bounded from above. Thus, there exists a constant $C > 0$ with $|K_{\sigma}(r - R_n(x), t - T_n(x))| \leq C$. Finally, the dominated convergence implies $\int_{\Omega} K_{\sigma}(r - R_n(x), t - T_n(x)) \, dx \rightarrow \int_{\Omega} K_{\sigma}(r - R(x), t - T(x))$ as $(R_n, T_n) \rightarrow (R,T)$. ∎

Using kernel functions for the computation of densities is not new. Popular candidates that have been proposed are (c.f. Figure 3.1):

Gaussian $K(x) = \varphi(x)$ where

$$\varphi(x) := \frac{1}{\sqrt{2\pi}} e^{-\frac{x^2}{2}} \tag{3.1}$$

B-splines $K(x) = \beta^m(x)$ where

$$\beta^m := \beta^{m-1} * \beta^0 \quad \text{and} \quad \beta^0(x) := \begin{cases} 1 & \text{if } |x| \leq \frac{1}{2}, \\ 0 & \text{if } |x| > \frac{1}{2}. \end{cases} \tag{3.2}$$

In particular the cubic B-spline β^2 is recently used [56, 57, 58].

A Trigonometric Function $K(x) = \psi(x)$ where

$$\psi(x) := \begin{cases} \frac{4}{3} \cos^4(\frac{\pi}{2} x) & \text{if } |x| \leq 1, \\ 0 & \text{if } |x| > 1. \end{cases} \tag{3.3}$$

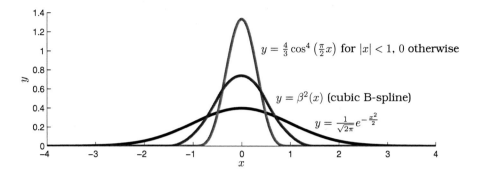

Figure 3.1: Some basic kernel functions

We introduced the concept of approximating densities to overcome several theoretical problems as well as to obtain explicit formulae for the computation of densities. Instead of approximating densities most authors make use of another method to compute densities - kernel density estimation. We will see this method involves some randomization. In order to present deterministic numerical methods in the following chapters, we will use only approximating densities. Nevertheless, kernel density estimation is an important method that is frequently used. Therefore, we give a short outline and reveal its connection to approximating densities.

3.2   Kernel Density Estimation

Kernel density estimation originates from statistics. It is devised to estimate the density of a random quantity without making prior assumptions, as, e.g., the density comes from a normal distribution. Therefore, kernel density estimation is a so-called non-parametric method. Furthermore, as common in statistics, in this section we call measurable functions *random variables*. The kernel density estimator is defined as follows:

Definition 3.4 (Kernel Density Estimator)

Let $X_1, X_2, \ldots, X_n : \Omega \to \mathbb{R}^k$ be *stochastically independent identically distributed random variables*, $X := (X_1, \ldots, X_n)$, and $K_\sigma : \mathbb{R}^k \to \mathbb{R}$ a *scaled kernel. The kernel density estimator is defined as*

$$\widehat{p}^\sigma(X; x) := \frac{1}{n} \sum_{j=1}^{n} K_\sigma(x - X_j) \quad \text{with} \quad \sigma > 0.$$

The random variables X_j model independent observations of a random quantity. For example, $X_1, X_2, \ldots, X_n$ might be the results of an experiment that was repeated n times.

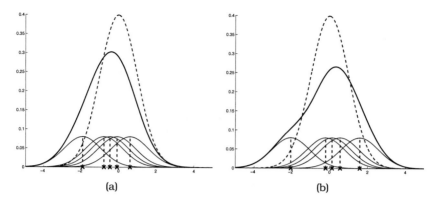

(a)   (b)

Figure 3.2: Two examples for kernel density estimates (solid) for the density normal distribution (dashed) based on five observations (cross) each.

However, if $X_1, X_2, \ldots, X_n$ are distributed by P_X with density p, it can be shown

$$\|\widehat{p}^{\sigma_n}(X; \cdot) - p\|_{L^1(\mathbb{R}^k)} \to 0 \quad \text{with probability one}$$

if $\sigma_n \to 0$ and $n\sigma_n \to \infty$ as $n \to \infty$ (cf. [51, §3.7] and references therein). Utilizing this concept for estimating the joint density p_{RT} of an image pair requires the construction of random variables from the images. Therefore, let $R, T \in \text{Img}(\Omega)$ and $X_1, X_2, \ldots, X_n : \Omega \to \Omega$ be stochastically independent identically uniform distributed random variables on Ω, i.e., $P_{X_j}(A) = \frac{|A|}{|\Omega|} = P(A)$ with P the normalized Lebesgue measure on Ω (cf. Definition 2.5). Hence, $X_j \sim P$, $j = 1, 2, \ldots, n$ and

$$(R(X_1), T(X_1)), (R(X_2), T(X_2)), \ldots, (R(X_n), T(X_n)) : \Omega \to \mathbb{R}^2$$

are stochastically independent identically distributed random variables. Furthermore, $(R(X_j), T(X_j))$, $j = 1, 2, \ldots, n$ are distributed by P_{RT}, i.e., $P_{R(X_j), T(X_j)} = P_{RT}$ since

$$\begin{aligned}
P_{R(X_j), T(X_j)}(A) &= P((R(X_j), T(X_j))^{-1}(A)) \\
&= P(X_j^{-1}((R, T)^{-1}(A))) \\
&= P_{X_j}((R, T)^{-1}(A)) \\
&= P((R, T)^{-1}(A)) \\
&= P_{RT}(A).
\end{aligned}$$

Therefore,

$$\widehat{p}_{RT}^{\sigma}(X; r, t) := \frac{1}{n} \sum_{j=1}^{n} K_\sigma(r - R(X_j), t - T(X_j)) \qquad (3.4)$$

is a kernel density estimator for the joint density.

In practice the construction of stochastically independent identically distributed random variables is done by randomly sampling the images [58, 62, 63, 66].

Recall that image registration is an optimization problem. For the design of a numerical method for the computation of a solution based on standard optimization methods, as, e.g., Newton's method or quasi Newton methods, we need a deterministic objective function. Due to this reason, many authors simply neglect the requirement of independent and identically distributed random variables by using a regular sampling of the images on a fixed grid [10, 12, 30, 40, 57].

However, there is a strong relation between approximating densities and the kernel density estimator (3.4). To this end we introduce the so-called *expectation.*

Definition 3.5 (Expectation)
Let $X : \Omega \to \mathbb{R}^k$ be a random variable with distribution P_X and $f : \mathbb{R}^k \to \mathbb{R}$ be integrable w.r.t. P_X. The expectation of f under X is defined as

$$E_X[f] := \int_{\mathbb{R}^k} f(x)\, \mathrm{d}P_X(x) = \int_{\Omega} f(X(\omega))\, \mathrm{d}P(\omega).$$

Now we are in a position to formulate the relation of the approximating density and the kernel density estimator. It turns out, the approximating density is nothing else than the expectation of the kernel density estimator.

Theorem 3.6 (Relation of $\widehat{p}_{RT}^{\,\sigma}$ and p_{RT}^{σ})
Let $R, T \in \mathrm{Img}(\Omega)$, $X_1, X_2, \ldots, X_n : \Omega \to \Omega$ be stochastically independent random variables identically distributed by P and $X := (X_1, X_2, \ldots, X_n)$. Then

$$E_X[\widehat{p}_{RT}^{\,\sigma}(X; r, t)] = p_{RT}^{\sigma}(r, t).$$

Proof.   Since $X_1, X_2, \ldots, X_n$ are stochastically independent identically distributed by P the distribution P_X of $X = (X_1, X_2, \ldots, X_n)$ is the n-times product measure of P, i.e., $X \sim P_X = \underbrace{P \otimes P \otimes \ldots \otimes P}_{n \text{ times}} = P^n$. This yields

$$E_X[\hat{p}_{RT}^\sigma(r,t)]$$

$$= \frac{1}{n} \sum_{j=1}^n \int_{\Omega^n} K_\sigma(r - R(x_j), t - T(x_j)) \, \mathrm{d}P^n(x_1, \ldots, x_n)$$

$$= \frac{1}{n} \sum_{j=1}^n \int_\Omega K_\sigma(r - R(x_j), t - T(x_j)) \underbrace{\int_{\Omega^{n-1}} \mathrm{d}P^{n-1}(x_1, .., x_{j-1}, x_{j+1}, .., x_n)}_{=1} \, \mathrm{d}P(x_j)$$

$$= \frac{1}{n} \sum_{j=1}^n \int_\Omega K_\sigma(r - R(x), t - T(x)) \, \mathrm{d}P(x)$$

$$= \int_\Omega K_\sigma(r - R(x), t - T(x)) \, \mathrm{d}P(x)$$

$$= \frac{1}{|\Omega|} \int_\Omega K_\sigma(r - R(x), t - T(x)) \, \mathrm{d}x$$

$$= p_{RT}^\sigma(r,t).$$

∎

In the following we will use approximating densities, since they allow for a theoretically sound deterministic setting. Nevertheless, the above theorem justifies their use even from a statistical point of view.

3.3 Approximating Mutual Information

The concept of mutual information originates from statistics based on random variables. So it is quite natural to use statistical methods for the registration based on mutual information, as well. Nevertheless, basically the registration problem is deterministic. Instead of adapting the registration problem to statistical concepts, here we adapt mutual information to image registration. Next, we define a functional that approximates the mutual information of two images. Therefore, we replace the densities of the images with the approximations given in the first section.

Definition 3.7 (Approximating Mutual Information)
Let $R, T \in \mathrm{Img}(\Omega)$ be images and p_{RT}^σ, p_R^σ, p_T^σ be the approximating densities given by Definition 3.1. We define the approximating mutual information as

$$\mathrm{MI}_\sigma[R,T] := \int_{\mathbb{R} \times \mathbb{R}} p_{RT}^\sigma(r,t) \, \log \frac{p_{RT}^\sigma(r,t)}{p_R^\sigma(r) \, p_T^\sigma(t)} \, \mathrm{d}(r,t).$$

Note that p_{RT}^σ is the density of $(R, T) + (\eta_1, \eta_2)$ with (η_1, η_2) stochastically independent from (R, T) and distributed with a density K_σ. Therefore, we have the identity

$$\mathrm{MI}_\sigma[R,T] = \mathrm{MI}[R + \eta_1, T + \eta_2].$$

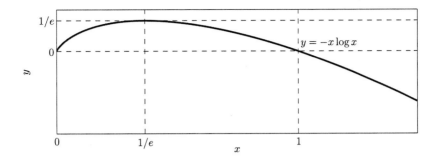

Figure 3.3: The "entropy function" $h(x) := -x \log x$ for $x > 0$ and $h(0) := 0$.

The so defined approximation has some nice properties. In particular $\mathrm{MI}_\sigma[R, T]$ is well-defined for nearly any kernels of interest, e.g., the kernels (3.1), (3.2), and (3.3) introduced in the previous section.

At the end of this section we give a theorem summarizing these properties. The theorem is based on approximating densities that are asymptotically bounded by a polynomial.

The following two lemmas will be helpful.

Lemma 3.8

Let $f \in L^1(\mathbb{R}^d)$ be a non-negative bounded function with $M := \sup_{x \in \mathbb{R}^d} f(x) < \infty$. If there exist constants $\alpha > d - 1$, $c \geq 0$, and $r_0 > 0$ such that $f(x) \leq c\|x\|^{-\alpha}$ whenever $\|x\| > r_0$ then

$$|f(x) \log f(x)| \leq \begin{cases} -c\|x\|^{-\alpha} \log c\|x\|^{-\alpha} & \text{if } \|x\| > r_0, \\ \max\{1/e, |M \log M|\} & \text{if } \|x\| \leq r_0 \end{cases}$$

and

$$\frac{r_0{}^d \, \pi^{d/2}}{\Gamma(\frac{d}{2} + 1)} \min\{0, -M \log M\} \leq - \int_{\mathbb{R}^d} f(x) \log f(x) \, dx$$

$$\leq \frac{2\alpha c^{\frac{d}{\alpha}} \, d \, \pi^{d/2}}{(\alpha + 1 - d)^2 \, \Gamma(\frac{d}{2} + 1)} + \frac{r_0{}^d \pi^{d/2}}{\Gamma(\frac{d}{2} + 1) \, e}$$

with the gamma function $\Gamma(x) = \int_0^\infty e^{-t} \, t^{x-1} \, dt$.

Proof.   The proof is based on the following properties of $h : \mathbb{R}_+ \to \mathbb{R}$ defined as $h(x) := -x \log x$ for $x > 0$ and $h(0) := 0$ (see Figure 3.3).

The function h is continuous, strictly monotonic increasing on $[0, 1/e]$, strictly monotonic decreasing on $[1/e, \infty)$, $h \geq 0$ on $[0, 1]$ with $h(0) = h(1) = 0$ and $h \leq 0$ on $[1, \infty)$. Furthermore, h assumes an unique maximum at $1/e$ with $h(1/e) = 1/e$.

Let $x \in \mathbb{R}^d$ with $\|x\| > r_0$. Without loss of generality we can assume $r_0 \geq (ce)^{\frac{1}{\alpha}}$ such that $0 \leq f(x) \leq c\|x\|^{-\alpha} \leq 1/e$. Since h is monotonic increasing and non-negative on $[0, 1/e]$ we have $0 \le h(f(x)) \le h(c||x||^{-\alpha})$ and therefore $|f(x) \log f(x)| \le c||x||^{-\alpha} \log c||x||^{-\alpha}$ if $||x|| > r_0$.

For $x \in \mathbb{R}^d$ with $||x|| \le r_0$ we have $0 \le f(x) \le M$. The continuity and monotony of h yields $|h(f(x))| \le \max\{1/e, |h(M)|\}$ if $||x|| \le r_0$.

Now we turn to the bounds for the integral. In order to use the estimates above we split the integral. Therefore, we define the sets $S := \{x : ||x|| \le r_0\}$ and $S^c := \mathbb{R}^d \setminus S$ such that

$$- \int_{\mathbb{R}^d} f(x) \log f(x) \, dx = - \int_S f(x) \log f(x) \, dx - \int_{S^c} f(x) \log f(x) \, dx.$$

If $M > 1/e$ we have $-M \log M \le -f(x) \log f(x) \le 1/e$ and $M \le 1/e$ yields $0 \le -f(x) \log f(x) \le -M \log M \le 1/e$ such that

$$|S| \min\{0, -M \log M\} \le - \int_S f(x) \log f(x) \, dx \le |S|/e.$$

Since S is a sphere with radius r_0 we have $|S| = \frac{r_0^d \pi^{d/2}}{\Gamma(d/2+1)}$.

The second integral over S^c is bounded by

$$0 \le - \int_{S^c} f(x) \log f(x) \, dx \le - \int_{S^c} c||x||^{-\alpha} \log c||x||^{-\alpha} \, dx.$$

$$= -\frac{d \pi^{d/2}}{\Gamma(\frac{d}{2}+1)} \int_{r_0}^{\infty} r^{d-1} \, cr^{-\alpha} \log cr^{-\alpha} \, dr.$$

$$\le -\frac{d \pi^{d/2}}{\Gamma(\frac{d}{2}+1)} \int_{(ce)^{\frac{1}{\alpha}}}^{\infty} r^{d-1} \, cr^{-\alpha} \log cr^{-\alpha} \, dr.$$

The substitution $r = (ce^t)^{\frac{1}{\alpha}}$, $dr = \frac{1}{\alpha}(ce^t)^{\frac{1}{\alpha}-1} ce^t \, dt = \frac{c^{\frac{1}{\alpha}}}{\alpha} e^{\frac{1}{\alpha}t} dt$ and $t = 1$ for $r = (ce)^{\frac{1}{\alpha}}$ yields

$$- \int_{(ce)^{\frac{1}{\alpha}}}^{\infty} r^{d-1} \, cr^{-\alpha} \log cr^{-\alpha} \, dr = -\frac{c^{\frac{d}{\alpha}}}{\alpha} \int_1^{\infty} e^{\frac{d-1}{\alpha}t} e^{-t} \log e^{-t} \, dt$$

$$= \frac{c^{\frac{d}{\alpha}}}{\alpha} \int_1^{\infty} e^{-\frac{\alpha+1-d}{\alpha}t} t \, dt$$

$$\le \frac{c^{\frac{d}{\alpha}}}{\alpha} \int_0^{\infty} e^{-\frac{\alpha+1-d}{\alpha}t} t \, dt$$

$$= \frac{\alpha c^{\frac{d}{\alpha}}}{(\alpha+1-d)^2} \underbrace{\int_0^{\infty} e^{-t} t \, dt}_{=\Gamma(2)=2!=2}$$

$$= \frac{2\alpha c^{\frac{d}{\alpha}}}{(\alpha+1-d)^2}$$

and therefore

$$0 \le - \int_{S^c} f(x) \log f(x) \, dx \le \frac{2\alpha c^{\frac{d}{\alpha}} \, d \, \pi^{d/2}}{(\alpha+1-d)^2 \, \Gamma(\frac{d}{2}+1)}.$$

Summing the bounds for the integrals asserts the lemma.   ∎

The above lemma asserts that the approximating mutual information is
bounded and therefore well-defined if the approximative density fulfills
the asymptotic growth condition of Lemma 3.8, since

$$
\begin{aligned}
\mathrm{MI}_\sigma[R, T] = &\int_{\mathbb{R} \times \mathbb{R}} p_{RT}^\sigma(r, t) \, \log p_{RT}^\sigma(r, t) \, \mathrm{d}(r, t) \\
&- \int_{\mathbb{R}} p_R^\sigma(r) \, \log p_R^\sigma(r) \, \mathrm{d}r - \int_{\mathbb{R}} p_T^\sigma(t) \, \log p_T^\sigma(t) \, \mathrm{d}t.
\end{aligned}
\tag{3.5}
$$

The asymptotic growth of the approximative density is basically a condition of the underlying kernel. The next lemma shows how the asymptotic
growth of a kernel transports to an approximating density.

Lemma 3.9
*Let $K \in L^1(\mathbb{R}^k)$ with $K(z) \leq c\|z\|^{-\alpha}$ when $\|z\| \geq z_0$. Furthermore let
$Z : \mathbb{R}^d \to \mathbb{R}^k$ be a function with $\|Z(x)\| \leq M$ for all $x \in \mathbb{R}^d$ and*

$$
p^\sigma(z) := \frac{1}{|\Omega|} \int_\Omega \frac{1}{\sigma^k} K\left(\frac{z - Z(x)}{\sigma}\right) \, \mathrm{d}x.
$$

Then $p^\sigma(z) \leq 2^\alpha \sigma^{\alpha-k} c \|z\|^{-\alpha}$ if $\|z\| \geq \max\{\sigma z_0 + M, 2M\}$.

Proof.   For all $z \in \mathbb{R}^k$ with $\|z\| \geq 2M$ holds

$$
\|z - Z(x)\| \geq \|z\| - \|Z(x)\| \geq \|z\| - M \geq \left(1 - \frac{M}{\|z\|}\right)\|z\| \geq \frac{1}{2}\|z\|.
$$

Furthermore, $\|z\| \geq \sigma z_0 + M$ implies $\left\|\frac{z - Z(x)}{\sigma}\right\| \geq \frac{1}{\sigma}(\|z\| - \|Z(x)\|) \geq \frac{1}{\sigma}(\|z\| - M) \geq z_0$ and therefore $\frac{1}{\sigma^k} K\left(\frac{z-Z(x)}{\sigma}\right) \leq \sigma^{\alpha-k} c \|z - Z(x)\|^{-\alpha} \leq c'\|z\|^{-\alpha}$ with $c' := 2^\alpha \sigma^{\alpha-k} c$. Hence

$$
p^\sigma(z) = \frac{1}{|\Omega|} \int_\Omega \frac{1}{\sigma^k} K\left(\frac{z - Z(x)}{\sigma}\right) \, \mathrm{d}x \leq \frac{1}{|\Omega|} \int_\Omega c'\|z\|^{-\alpha} \, \mathrm{d}x = c'\|z\|^{-\alpha}
$$

whenever $\|z\| \geq \max\{\sigma z_0 + M, 2M\}$.   ∎

Examples for polynomial asymptotically bounded functions are the kernels (3.1), (3.2), and (3.3).

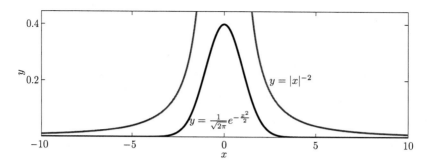

Figure 3.4: An asymptotic bound of the Gaussian

Lemma 3.10 (Polynomial Asymptotically Bounded Kernels)
Let φ be the Gaussian (3.1), β^m the B-spline (3.2), and ψ be the trigono-metric kernel function (3.1). Furthermore, for $x = (x_1, x_2, \ldots, x_k) \in \mathbb{R}^k$ we define their tensor-products

$$K^{\text{GAUSS}}(x) := \varphi(x_1) \cdot \varphi(x_2) \cdots \varphi(x_k) = \left(\frac{1}{2\pi}\right)^{\frac{d-1}{2}} \varphi(\|x\|),$$

$$K^{\text{B-SPLINE}}(x) := \beta^m(x_1) \cdot \beta^m(x_2) \cdots \beta^m(x_k),$$

$$K^{\text{COS}}(x) := \psi(x_1) \cdot \psi(x_2) \cdots \psi(x_k).$$

The functions K^{GAUSS}, $K^{\text{B-SPLINE}}$, and K^{COS} are polynomial asymptotically bounded, i.e., for $K \in \{K^{\text{GAUSS}}, K^{\text{B-SPLINE}}, K^{\text{COS}}\}$ exist constants $\alpha, c, x_0 > 0$ such that $K(x) \leq c \|x\|^{-\alpha}$ for all $x \in \mathbb{R}^k$ with $\|x\| \geq x_0$.

Proof. The Gaussian φ decays exponentially and therefore faster than any polynomial. Hence, $K^{\text{GAUSS}}(x) = (2\pi)^{-\frac{d-1}{2}} \varphi(\|x\|)$ is polynomial asymptotically bounded.

The B-spline β^m and the trigonometric kernel ψ are compactly supported. Therefore, their k-fold tensor products $K^{\text{B-SPLINE}}$ are K^{COS} are compactly supported as well and hence asymptotically bounded by any polynomial.
∎

Thus, working with these kernels of practical interest fits into our framework for approximating densities.

Finally, we summarize properties of the approximative mutual information.

Theorem 3.11 (Properties of MI_σ)

Let $R, T \in \mathrm{Img}(\Omega)$ be images and MI_σ based on approximating densities p_{RT}^σ, p_R^σ, p_T^σ with a kernel K. Then the following statements hold:

a) $\mathrm{MI}_\sigma[R, T] = \mathrm{MI}_\sigma[T, R]$ *(symmetry)*

b) $0 \le \mathrm{MI}_\sigma[R, T]$ with equality if and only if R and T are stochastically independent. *(positive definiteness)*

Additionally, if K fulfills the growth condition of Lemma 3.8 and the images are bounded by $M > 0$, i.e., $\max\{\sup_{x \in \Omega} |R(x)|, \sup_{x \in \Omega} |T(x)|\} \le M$, then holds:

c) There exists a constant C depending on the kernel K and the scaling σ, such that $\mathrm{MI}_\sigma[R, T] \le C$. *(boundedness)*

d) Let $(R_n, T_n) \subset \mathrm{Img}(\Omega)^2$ be a sequence of image pairs with intensities bounded by M, that is $\max\{\sup_{x \in \Omega} |R_n(x)|, \sup_{x \in \Omega} |T_n(x)|\} \le M$ for all n. Then

$$\mathrm{MI}_\sigma[R_n, T_n] \to \mathrm{MI}_\sigma[R, T] \quad \text{as} \quad (R_n, T_n) \to (R, T).$$

(continuity)

Proof. a) From the definition of p_{RT}^σ follows $p_{RT}^\sigma(r, t) = p_{TR}^\sigma(t, r)$ and therefore $\mathrm{MI}_\sigma[R, T] = \mathrm{MI}_\sigma[T, R]$.

b) For stochastically independent images R and T the joint distribution degenerates to the product of its marginals, that is $P_{RT} = P_R \otimes P_T$. Thus

$$
\begin{aligned}
p_{RT}^\sigma(r, t) &= \frac{1}{|\Omega|} \int_\Omega K_\sigma(r - R(x)) \, K_\sigma(t - T(x)) \, \mathrm{d}x \\
&= \int_{\mathbb{R} \times \mathbb{R}} K_\sigma(r - \varrho) \, K_\sigma(t - \vartheta) \, \mathrm{d}P_{RT}(\varrho, \vartheta) \\
&= \int_{\mathbb{R} \times \mathbb{R}} K_\sigma(r - \varrho) \, K_\sigma(t - \vartheta) \, \mathrm{d}P_R(\varrho) \, \mathrm{d}P_T(\vartheta) \\
&= \left(\int_{\mathbb{R}} K_\sigma(r - \varrho) \, \mathrm{d}P_R(\varrho) \right) \left(\int_{\mathbb{R}} K_\sigma(t - \vartheta) \, \mathrm{d}P_T(\varrho) \right) \\
&= \left(\frac{1}{|\Omega|} \int_\Omega K_\sigma(r - R(x)) \, \mathrm{d}x \right) \left(\frac{1}{|\Omega|} \int_\Omega K_\sigma(t - T(x)) \, \mathrm{d}x \right) \\
&= p_R^\sigma(r) \, p_T^\sigma(t)
\end{aligned}
$$

Then b) follows in the same manner as in the proof of Theorem 2.18b.

c) Expanding the integral we find

$$\mathrm{MI}_\sigma = \int_{\mathbb{R} \times \mathbb{R}} p_{RT}^\sigma \log p_{RT}^\sigma \, \mathrm{d}(r, t) - \int_{\mathbb{R}} p_R^\sigma \log p_R^\sigma \, \mathrm{d}r - \int_{\mathbb{R}} p_T^\sigma \log p_T^\sigma \, \mathrm{d}t.$$

Lemma 3.9 asserts that p_{RT}^σ, p_R^σ, p_T^σ are asymptotically bounded by a polynomial. Furthermore, from the definition of the approximating densities we have $p_{RT}^\sigma(r,t) \le \frac{1}{\sigma^2}K(0)^2$ and $p_R^\sigma(r) \le \frac{1}{\sigma}K(0)$, $p_T^\sigma(r) \le \frac{1}{\sigma}K(0)$ for all r,t. Thus, we can apply Lemma 3.8 to the three integrals separately and we obtain bounds only depending on the kernel K. Summing the bounds yields the statement.

d) As in the proof of c) we expand the integral and treat them separately. We only show

$$\int_{\mathbb{R}} p_{R_n}^\sigma(r) \log p_{R_n}^\sigma(r) dr \quad \to \quad \int_{\mathbb{R}} p_R^\sigma(r) \log p_R^\sigma(r) dr.$$

The proofs for $\int p_{R_n T_n}^\sigma \log p_{R_n T_n}^\sigma \to \int p_{RT}^\sigma \log p_{RT}^\sigma$ and $\int p_{T_n}^\sigma \log p_{T_n}^\sigma \to \int p_T^\sigma \log p_T^\sigma$ are in the same line.

By assumption the kernel K of $p_{R_n}^\sigma$ fulfills the assumptions of Lemma 3.8 and therefore $c, \alpha > 0$ exists with $K(r) \le c|r|^{-\alpha}$ if $|r| > r_0$. Thus, Lemma 3.9 asserts $p_{R_n}^\sigma(r) \le c'|r|^{-\alpha}$ if $|r| \ge \max\{\sigma r_0 + M, 2M\} =: r_0'$ with $c' = 2^\alpha \sigma^{\alpha-1}$ and from Lemma 3.8 we have $\left|p_{R_n}^\sigma(r) \log p_{R_n}^\sigma(r)\right| \le F(r)$ with $F(r) = -c'|r|^{-\alpha} \log c'|r|^{-\alpha}$ if $|r| \ge r_0'$ and $F(r) = \max\{1/e, |M \log M|\}$ if $|r| \le r_0'$. By assumption $|R_n(x)| \le M$ for all $n \in \mathbb{N}$ such that $\left|p_{R_n}^\sigma(r) \log p_{R_n}^\sigma(r)\right| \le F(r)$ holds for all n. From Lemma 3.3 and the continuity of the log we have $p_{R_n}^\sigma \log p_{R_n}^\sigma \to p_R^\sigma \log p_R^\sigma$. Furthermore F is integrable on $\mathbb{R}$ such that the dominated convergence implies the convergence of the integrals $\int_{\mathbb{R}} p_{R_n}^\sigma \log p_{R_n}^\sigma \, dr \to \int_{\mathbb{R}} p_R^\sigma \log p_R^\sigma \, dr$.   ∎

Chapter 4

PDE Based Methods

In this chapter we discuss PDE (partial differential equation) based methods for multi-modal image registration based on mutual information. We will establish time-dependent (parabolic) PDE's that evolved over time yield the alignment of two images. Therefore, the PDE models a continuous process that can be interpreted as motion. As a result, we finally obtain not only a transformation that aligns two images, but also a motion pathway from the template to the reference image. This interpretation also reveals the strong relation of image registration and optical flow estimation [6].

4.1 Approaches

Recalling from Chapter 1 the image registration problem reads

$$\mathcal{D}[u] + \alpha \mathcal{S}[u] \overset{u}{\to} \min$$

with a distance measure $\mathcal{D}[u] := \mathcal{D}[R, T \circ (\mathrm{id} - u)]$ and a smoother $\mathcal{S}$ (cf. Problem 1.4 on page 9). As smoother here we consider the diffusive, curvature, and elastic smoother. As distance measure we use the negative mutual information. Therefore, we treat the minimization problem

$$\mathcal{D}^{\mathrm{MI}}[u] + \alpha \mathcal{S}[u] \overset{u}{\to} \min \quad \text{with} \quad \mathcal{D}^{\mathrm{MI}}[u] := -\mathrm{MI}[R, T \circ (\mathrm{id} - u)]. \quad (4.1)$$

As we saw in Chapter 2, dealing with mutual information directly involves several theoretical problems. In particular, we saw that densities may not exist and the mutual information functional is not well-defined. To overcome these problems in Chapter 3 we introduced the approximative mutual information MI_σ (cf. Definition 3.7 on page 50). In a second approach we consider the minimization problem

$$\mathcal{D}^{\mathrm{MI}_\sigma}[u] + \alpha \mathcal{S}[u] \overset{u}{\to} \min \quad \text{with} \quad \mathcal{D}^{\mathrm{MI}_\sigma}[u] := -\mathrm{MI}_\sigma[R, T \circ (\mathrm{id} - u)]. \quad (4.2)$$

Using such a functional for image registration based on mutual information was proposed by Hermosillo in 2002 [30]. His approach is based on the particular choice of the Gaussian as kernel for approximating densities and mutual information respectively. Furthermore, the interpretation of the density approximation is different from ours. The density approximation in [30] is seen as a kind of continuous kernel density estimator (cf. §3.2). However, in the following we refer to (4.1) as *direct approach* and to (4.2) as *approximative approach*.

Applying the calculus of variations we will derive necessary conditions for minimizers. These conditions are the so-called *Euler-Lagrange equations*. The Euler-Lagrange equations correspond to the well-known condition from optimization that a minimizer must be a zero of the gradient.

Next we shortly review the computation of the Euler-Lagrange equations and determine building blocks for the registration problem. To keep things simple, in the following we always consider continuous differentiable functions of a linear space $V \subseteq C^m(\overline{\Omega}; \mathbb{R}^d)$ and $\Omega = (0,1)^d$. Nevertheless, the following results hold also in general true for less smooth functions in the Sobolev-space $H^m(\Omega)$ and Ω a domain with sufficiently smooth boundary (Lipschitz) instead of functions in $C^m(\overline{\Omega})$. However, the above assumptions provide everything we need. For more details see, e.g., the textbooks [1, 16]. Furthermore, we have need for the normal field of $\partial\Omega$. In the following, we denote the outward pointing normal field by $n : \partial\Omega \to \mathbb{R}^d$.

4.2 Euler-Lagrange Equations

Let $\Omega = (0,1)^d$, $V \subseteq C^m(\overline{\Omega})$ be a linear space and $\|\cdot\|_{L^2(\Omega)}$ the L^2 norm on Ω, i.e.,

$$\|u\|_{L^2(\Omega)} := \left(\int_\Omega \|u\|^2 \, \mathrm{d}x \right)^{\frac{1}{2}}.$$

The image registration problem is to find a (local) minimizer of a functional $\mathcal{J} : V \to \mathbb{R}$ (cf. Problem 1.3),

$$\mathcal{J}[u] := \mathcal{D}[u] + \alpha \mathcal{S}[u]$$

with a distance measure $\mathcal{D}$ and a smoother $\mathcal{S}$. That is, we want to compute a function $u \in V$ for which an $\varepsilon > 0$ exists such that

$$\mathcal{J}[u] \leq \mathcal{J}[v] \quad \text{holds for all } v \text{ with } \|u - v\|_{L^2(\Omega)} < \varepsilon.$$

Defining $\phi(\tau) := \mathcal{J}[u + \tau v]$ for a minimizer u and an arbitrary fixed v with $\|v\|_{L^2(\Omega)} = 1$ then $\phi(0)$ is a (local) minimum and therefore $\tau = 0$ a (local) minimizer for ϕ. For differentiable ϕ and $\mathcal{J}$ respectively $\tau = 0$ must be a zero of the first order derivative, thus

$$0 = \phi'(0) = \frac{\mathrm{d}\mathcal{J}[u + \tau v]}{\mathrm{d}\tau}\bigg|_{\tau=0}. \tag{4.3}$$

Note that $\phi'(0)$ is the directional derivative of $\mathcal{J}$ at u in direction v. This directional derivative is the so-called *Gateâux derivative* and we define

$$d\mathcal{J}[u;v] := \frac{\mathrm{d}}{\mathrm{d}\tau}\mathcal{J}[u + \tau v]\bigg|_{\tau=0}. \tag{4.4}$$

Since v was arbitrarily chosen, (4.3) holds for all $v \in V$. Thus a necessary condition for a minimizer u is

$$d\mathcal{J}[u;v] = 0 \quad \text{for all } v \in V. \tag{4.5}$$

This equation is called Euler-Lagrange equation (ELE) in its weak form. Note that

$$\frac{\mathrm{d}\mathcal{J}[u + \tau v]}{\mathrm{d}\tau} = \frac{\mathrm{d}\mathcal{D}[u + \tau v]}{\mathrm{d}\tau} + \alpha\frac{\mathrm{d}\mathcal{S}[u + \tau v]}{\mathrm{d}\tau}$$

and therefore $d\mathcal{J}[u;v] = d\mathcal{D}[u;v] + \alpha\, d\mathcal{S}[u;v]$.

To outline the general setting for the classical formulation of Euler-Lagrange equations as boundary value problems we introduce the L^2 inner product. For functions $u, v : \mathbb{R}^d \to \mathbb{R}^d$ the L^2 inner product is defined as

$$\langle u, v\rangle_{L^2(\Omega)} := \int_\Omega u \cdot v \; \mathrm{d}x \quad \text{with} \quad u \cdot v := \sum_{\ell=1}^d u_\ell v_\ell$$

where $u \cdot v$ is the standard or Euclidean inner product on $\mathbb{R}^d$. In the following, we also write $\langle u, v\rangle$ for $u \cdot v$. Then the Gateâux derivatives of the distance measure and the smoother can be written as

$$d\mathcal{D}[u;v] = \langle f(u), v\rangle_{L^2(\Omega)} \quad \text{and} \quad d\mathcal{S}[u;v] = \langle \mathcal{L}u, \mathcal{L}v\rangle_{L^2(\Omega)}.$$

with a non-linear function f, a so-called *force*, and a linear differential operator $\mathcal{L} : C^m(\overline{\Omega}) \to C^0(\overline{\Omega})$. Thereby $f : \mathbb{R}^d \times \mathbb{R}^d \to \mathbb{R}^d$ and $f(u)$ is short-hand notation of $f(\cdot, u(\cdot))$.

For sufficiently smooth functions $u, v \in C^{2m}(\overline{\Omega})$ we can apply integration by parts to the Gateâux derivative of the smoother and we obtain

$$d\mathcal{S}[u;v] = \langle \mathcal{A}u, v\rangle_{L^2(\Omega)} \quad + \quad \text{boundary integrals}$$

with a linear differential operator $\mathcal{A} : V \subseteq C^{2m}(\overline{\Omega}) \to C^0(\overline{\Omega})$, given by $\mathcal{A} := \mathcal{L}^*\mathcal{L}$ with $\mathcal{L}^*$ the *adjoint* operator of $\mathcal{L}$ w.r.t. the L^2 inner product. The adjoint operator $\mathcal{L}^*$ of $\mathcal{L}$ is uniquely determined by the relation

$$\langle u, \mathcal{L}v\rangle_{L^2(\Omega)} = \langle \mathcal{L}^*u, v\rangle_{L^2(\Omega)} \quad \text{for all } u \in C^m(\Omega) \text{ and } v \in C_c^\infty(\Omega).$$

Thus, the Euler-Lagrange equation (4.5) reads

$$d\mathcal{J}[u;v] = \langle f(u) + \mathcal{A}u, v\rangle_{L^2(\Omega)} + \text{boundary integrals} = 0$$

for all $v \in V$. Based on the following well-known lemma, this is only valid for all v if $f(u) + \mathscr{A}u = 0$ and u fulfills boundary conditions that make the boundary integrals vanish, too.

Lemma 4.1

Let $\Omega \subset \mathbb{R}^d$ be a domain and $f \in C^0(\overline{\Omega})$. Then

a) $\displaystyle \int_\Omega f(x)\,g(x)\,\mathrm{d}x = 0$ for all $g \in C_c^\infty(\Omega)$ $\Leftrightarrow$ $f = 0$ on $\overline{\Omega}$,

b) $\displaystyle \int_{\partial \Omega} f(x)\,g(x)\,\mathrm{d}S(x) = 0$ for all $g \in C^\infty(\overline{\Omega})$ $\Leftrightarrow$ $f = 0$ on $\partial \Omega$.

Proof.   We only show the implications "$\Rightarrow$". a) Assume $\int_\Omega fg\,\mathrm{d}x = 0$ for all $g \in C_c^0(\Omega)$ and a $x_0 \in \Omega$ exists with $f(x_0) \neq 0$. Without loss of generality we can assume $f(x_0) > 0$. Then the continuity of f implies that f is also non-negative in a neighborhood of x_0. That is, an $\varepsilon > 0$ exists such that $f(x) > 0$ for all $x \in B_\varepsilon(x_0) = \{x : \|x - x_0\| \leq \varepsilon\} \subset \Omega$. We can assume $\overline{B_\varepsilon(x_0)} \subset \Omega$ since $\overline{B_{\varepsilon'}(x_0)} \subset B_\varepsilon(x_0)$ for all $\varepsilon' < \varepsilon$. Now we can find a function $g \in C_c^\infty(\Omega)$ with $g > 0$ on $B_\varepsilon(x_0)$ and $g = 0$ on $\Omega \setminus B_\varepsilon(x_0)$, e.g., $g(x) := g_0(\|x - x_0\|/\epsilon)$ with $g_0(t) := \exp(-1/(1 - t^2))$ for $|t| < 1$ and otherwise 0.

Then $fg > 0$ on $B_\varepsilon(x_0)$ and $\int_\Omega fg\,\mathrm{d}x = \int_{B_\varepsilon(x_0)} fg\,\mathrm{d}x > 0$ is a contradiction to $\int_\Omega fg\,\mathrm{d}x = 0$. Thus, $f(x) = 0$ must hold for all $x \in \Omega$. Finally, the continuity of f implies $f = 0$ on $\partial \Omega$ such that $f(x) = 0$ holds for all $x \in \overline{\Omega}$.
b) Analogously to the proof of a) we assume without loss of generality $f(x_0) > 0$ for a $x_0 \in \partial \Omega$ such that $\varepsilon > 0$ exists with $f(x) > 0$ for all $x \in B_\varepsilon(x_0) \cap \overline{\Omega}$. Then we can find $g : U \to \mathbb{R} \in C^\infty(\overline{\Omega})$ defined on an open set $U \supset \overline{\Omega}$ with $g(x) > 0$ on $B_\varepsilon(x_0)$ and $g(x) = 0$ on $U \setminus B_\varepsilon(x_0)$. Thus $fg > 0$ on $\partial \Omega \cap B_\varepsilon(x_0)$ and therefore $\int_{\partial \Omega} f\,g\,\mathrm{d}S = \int_{\partial \Omega \cap B_\varepsilon(x_0)} f\,g\,\mathrm{d}S > 0$ is a contradiction to $\int_{\partial \Omega} f\,g\,\mathrm{d}S = 0$.   ∎

Summarizing, a function u is a solution to the Euler-Lagrange equation (4.5), if and only if, it is a solution to the boundary value problem (BVP)

$$f(u) + \mathscr{A}u = 0 \quad + \quad \text{boundary conditions.}$$

This is the classical formulation of the Euler-Lagrange equations.
As we have seen, the Euler-Lagrange equations have two building blocks. First, the force f depending on the distance measure. Second, the differential operator $\mathscr{A}$ depending on the smoother. The several registration approaches are combinations of these building blocks. In particular the smoother determines the type of the boundary problem to be solved for the registration. Next, we compute the Gateâux derivatives and therefore forces of the distance measures $\mathcal{D}^{\mathrm{MI}_\sigma}$ and $\mathcal{D}^{\mathrm{MI}}$. Subsequently, we compute

the Gateâux derivatives of the smoothers and derive the classical formulation of the Euler-Lagrange equations as boundary value problems.

4.2.1 Gateâux Derivatives of $\mathcal{D}^{\mathrm{MI}}$ and $\mathcal{D}^{\mathrm{MI}_\sigma}$

We start with the direct approach. The following lemma gives the Gateâux
derivative of $\mathcal{D}^{\mathrm{MI}}$.

Lemma 4.2 (Gateâux Derivatives of $\mathcal{D}^{\mathrm{MI}}$)

*Let $u, v : \mathbb{R}^d \to \mathbb{R}^d \in C^0(\overline{\Omega}; \mathbb{R}^d)$ and $R \in \mathrm{Img}(\Omega)$, $T \in \mathrm{Img}(\Omega) \cap C^1(\Omega)$ be
images and $T_{u+\tau v} := T \circ (\mathrm{id} - (u + \tau v))$ with $\tau \in \mathbb{R}$.
If $\tau_0 > 0$ exists such that $(R, T_{u+\tau v})$ is absolutely continuous with joint
density $p_{RT_{u+\tau v}} \in C^1(\mathbb{R}^2)$ and $p_{RT_{u+\tau v}}$ is differentiable w.r.t. τ, whenever
$|\tau| < \tau_0$, then $\mathcal{D}^{\mathrm{MI}}$ is Gateâux differentiable at u in direction v and its
derivative is given by*

$$d\mathcal{D}^{\mathrm{MI}}[u; v] = \left\langle f^{\mathrm{MI}}(u), v \right\rangle_{L^2(\Omega)}$$

with force

$$f^{\mathrm{MI}}(x, u(x)) := L_{RT_u}\Big(R(x), T(x - u(x)) \Big) \, \nabla T(x - u(x))$$

where

$$L_{RT_u}(r, t) := \frac{1}{|\Omega|} \left(\frac{\partial_2 p_{RT_u}(r, t)}{p_{RT_u}(r, t)} - \frac{p'_{T_u}(t)}{p_{T_u}(t)} \right).$$

Proof. From (2.19) on page (28) we have the representation

$$\mathcal{D}^{\mathrm{MI}}[u + \tau v] = -\mathrm{MI}[R, T_{u+\tau v}] = -\frac{1}{|\Omega|} \int_\Omega \log \frac{p_{RT_{u+\tau v}}(R, T_{u+\tau v})}{p_R(R)\, p_{T_{u+\tau v}}(T_{u+\tau v})} \; \mathrm{d}x.$$

Differentiating w.r.t. τ yields

$$\frac{\mathrm{d}\mathcal{D}^{\mathrm{MI}}[u + \tau v]}{\mathrm{d}\tau} = -\frac{\mathrm{d}}{\mathrm{d}\tau} \frac{1}{|\Omega|} \int_\Omega \log \frac{p_{RT_{u+\tau v}}(R, T_{u+\tau v})}{p_R(R)\, p_{T_{u+\tau v}}(T_{u+\tau v})} \; \mathrm{d}x$$

$$= \frac{1}{|\Omega|} \int_\Omega \frac{\frac{\mathrm{d}}{\mathrm{d}\tau} p_{T_{u+\tau v}}(T_{u+\tau v})}{p_{T_{u+\tau v}}(T_{u+\tau v})} \; \mathrm{d}x - \frac{1}{|\Omega|} \int_\Omega \frac{\frac{\mathrm{d}}{\mathrm{d}\tau} p_{RT_{u+\tau v}}(R, T_{u+\tau v})}{p_{RT_{u+\tau v}}(R, T_{u+\tau v})} \; \mathrm{d}x$$

For the first integral we have

$$\frac{1}{|\Omega|} \int_\Omega \frac{\frac{\mathrm{d}}{\mathrm{d}\tau} p_{T_{u+\tau v}}(T_{u+\tau v})}{p_{T_{u+\tau v}}(T_{u+\tau v})} \, \mathrm{d}x$$

$$= \frac{1}{|\Omega|} \int_\Omega \frac{-p'_{T_{u+\tau v}}(T_{u+\tau v}) \, \nabla T_{u+\tau v} \cdot v - \frac{\partial p_{T_{u+\tau v}}}{\partial \tau}(T_{u+\tau v})}{p_{T_{u+\tau v}}(T_{u+\tau v})} \, \mathrm{d}x$$

$$= -\frac{1}{|\Omega|} \int_\Omega \frac{p'_{T_{u+\tau v}}(T_{u+\tau v}) \, \nabla T_{u+\tau v} \cdot v}{p_{T_{u+\tau v}}(T_{u+\tau v})} \, \mathrm{d}x - \frac{1}{|\Omega|} \int_\Omega \frac{\frac{\partial p_{T_{u+\tau v}}}{\partial \tau}(T_{u+\tau v})}{p_{T_{u+\tau v}}(T_{u+\tau v})} \, \mathrm{d}x$$

$$= -\frac{1}{|\Omega|} \int_\Omega \frac{p'_{T_{u+\tau v}}(T_{u+\tau v})}{p_{T_{u+\tau v}}(T_{u+\tau v})} \nabla T_{u+\tau v} \cdot v \, \mathrm{d}x$$

since

$$\frac{1}{|\Omega|} \int_\Omega \frac{\frac{\partial p_{T_{u+\tau v}}}{\partial \tau}(T_{u+\tau v})}{p_{T_{u+\tau v}}(T_{u+\tau v})} \, \mathrm{d}x = \int_{\mathrm{R}} \frac{\frac{\partial p_{T_{u+\tau v}}}{\partial \tau}(t)}{p_{T_{u+\tau v}}(t)} \, \mathrm{d}P_{T_{u+\tau v}}(t) = \int_{\mathrm{R}} p_{T_{u+\tau v}}(t) \frac{\frac{\partial p_{T_{u+\tau v}}}{\partial \tau}(t)}{p_{T_{u+\tau v}}(t)} \, \mathrm{d}t$$

$$= \int_{\mathrm{R}} \frac{\partial p_{T_{u+\tau v}}}{\partial \tau}(t) \, \mathrm{d}t = \frac{\mathrm{d}}{\mathrm{d}\tau} \underbrace{\int_{\mathrm{R}} p_{T_{u+\tau v}}(t) \, \mathrm{d}t}_{=1} = 0.$$

An analogous computation shows

$$\frac{1}{|\Omega|} \int_\Omega \frac{\frac{\mathrm{d}}{\mathrm{d}\tau} p_{RT_{u+\tau v}}(R, T_{u+\tau v})}{p_{R, T_{u+\tau v}}(R, T_{u+\tau v})} \, \mathrm{d}x = -\frac{1}{|\Omega|} \int_\Omega \frac{\partial_2 p_{RT_u}(R, T_u)}{p_{R, T_u}(R, T_u)} \nabla T_u \cdot v \, \mathrm{d}x.$$

Setting $\tau = 0$ and summing both integrals we finally obtain

$$d\mathcal{D}^{\mathrm{MI}}[u; v] = \int_\Omega \left(\frac{\partial_2 p_{RT_u}(R, T_u)}{p_{R, T_u}(R, T_u)} - \frac{p'_{T_u}(T_u)}{p_{T_u}(T_u)} \right) \nabla T_u \cdot v \, \mathrm{d}x.$$

∎

As already mentioned there are theoretical difficulties with the well-posedness of $\mathcal{D}^{\mathrm{MI}}$ and therefore its derivative $d\mathcal{D}^{\mathrm{MI}}$. These difficulties are reflected by the numerous assumptions made in the above lemma. We can hardly ensure the existence or even the required differentiability of the densities occuring in the force f^{MI} for a reliable class of images and transformations by theoretical arguments. Nevertheless, the strength of the direct approach is that we are free of any particular approximations and estimates of the densities. From a practical point of view we replace the occuring densities with suitable approximations, e.g., the approximating densities given by Definition 3.1 or the kernel density estimates given in §3.2. For example, using the approximative densities results

$$f^{\mathrm{MI}}(x, u(x)) \approx L^\sigma_{RT_u}\left(R(x), T(x - u(x)) \right) \nabla T(x - u(x)) \tag{4.6}$$

with

$$L^\sigma_{RT_u}(r,t) := \frac{1}{|\Omega|} \left(\frac{\partial_2 p^\sigma_{RT_u}(r,t)}{p^\sigma_{RT_u}(r,t)} - \frac{p^{\sigma\prime}_{T_u}(t)}{p^\sigma_{T_u}(t)} \right).$$

Keeping this in mind, we use the force f^{MI} in the following with the implicit understanding of a well-defined continuous function.

Now we turn to the approximating approach. The following lemma gives the Gateâux Derivative of $\mathcal{D}^{\mathrm{MI}_\sigma}$. The lemma is in line with the Gateâux derivative presented in [30] in the particular case of approximating densities based on the Gaussian.

Lemma 4.3 (Gateâux Derivatives of $\mathcal{D}^{\mathrm{MI}_\sigma}$)

Let $u \in C^0(\overline{\Omega}; \mathbb{R}^d)$ and $R, T \in \mathrm{Img}(\Omega)$ be images. Furthermore let $p^\sigma_{RT_u}$, p^σ_R, and $p^\sigma_{T_u}$ be the approximative densities of R and $T \circ (\mathrm{id} - u)$ based on a kernel K_σ as given by Definition 3.1. Then

$$d\mathcal{D}^{\mathrm{MI}_\sigma}[u; v] = \left\langle f^{\mathrm{MI}_\sigma}(u), v \right\rangle_{L^2(\Omega)}$$

with force

$$f^{\mathrm{MI}_\sigma}(x, u(x)) := \left(K_\sigma * L^\sigma_{RT_u} \right)\!\left(R(x), T(x - u(x)) \right) \nabla T(x - u(x))$$

where

$$L^\sigma_{RT_u}(r,t) := \frac{1}{|\Omega|} \left(\frac{\partial_2 p^\sigma_{RT_u}(r,t)}{p^\sigma_{RT_u}(r,t)} - \frac{p^{\sigma\prime}_{T_u}(t)}{p^\sigma_{T_u}(t)} \right).$$

Proof. To keep notation short we define $T_u(x) := T(x - u(x))$ and $\nabla T_u(x) := (\nabla T)(x - u(x))$. Furthermore, let the approximating densities, in addition, depend on τ in the manner

$$p^\sigma_{RT_u}(r,t,\tau) := \frac{1}{|\Omega|} \int_\Omega K_\sigma(r - R(x)) K_\sigma\!\left(t - T(x - (u + \tau v)(x)) \right)\, \mathrm{d}x$$

and $p^\sigma_{T_u}(t, \tau) := \int_{\mathbb{R}} p^\sigma_{RT_u}(r,t,\tau)\, \mathrm{d}r$. Then

$$
\begin{aligned}
\frac{\mathrm{d}\mathcal{D}^{\mathrm{MI}_\sigma}[u + \tau v]}{\mathrm{d}\tau} &= -\frac{\mathrm{dMI}_\sigma[u + \tau v]}{\mathrm{d}\tau} \\
&= -\frac{\mathrm{d}}{\mathrm{d}\tau} \int_{\mathbb{R}^2} p^\sigma_{RT_u}(r,t,\tau)\, \log \frac{p^\sigma_{RT_u}(r,t,\tau)}{p^\sigma_R(r)\, p^\sigma_{T_u}(t,\tau)}\, \mathrm{d}(r,t) \\
&= \int_{\mathbb{R}^2} \frac{\mathrm{d}p^\sigma_{RT_u}(r,t,\tau)}{\mathrm{d}\tau} \left(1 - \log \frac{p^\sigma_{RT_u}(r,t,\tau)}{p^\sigma_R(r)\, p^\sigma_{T_u}(t,\tau)} \right) \mathrm{d}(r,t) \\
&\quad - \int_{\mathbb{R}^2} \frac{p^\sigma_{RT_u}(r,t,\tau)}{p^\sigma_{T_u}(t,\tau)} \frac{\mathrm{d}p^\sigma_{T_u}(t,\tau)}{\mathrm{d}\tau}\, \mathrm{d}(r,t) \\
&= \int_{\mathbb{R}^2} \frac{\mathrm{d}p^\sigma_{RT_u}(r,t,\tau)}{\mathrm{d}\tau} \left(1 - \log \frac{p^\sigma_{RT_u}(r,t)}{p^\sigma_R(r)\, p^\sigma_{T_u}(t)} \right) \mathrm{d}(r,t)
\end{aligned}
$$

since

$$\int_{\mathbb{R}^2} \frac{p^\sigma_{RT_u}(r,t,\tau)}{p^\sigma_{T_u}(t,\tau)} \frac{\mathrm{d}p^\sigma_{T_u}(t,\tau)}{\mathrm{d}\tau} \, \mathrm{d}(r,t) = \int_{\mathbb{R}} \frac{\mathrm{d}p^\sigma_{T_u}(t,\tau)}{\mathrm{d}\tau} \, \mathrm{d}t = \frac{\mathrm{d}}{\mathrm{d}\tau} \underbrace{\int_{\mathbb{R}} p^\sigma_{T_u}(t,\tau) \, \mathrm{d}t}_{=1} = 0.$$

Setting $\tau = 0$, the derivative of $p^\sigma_{RT_u}(r,t,\tau)$ w.r.t. τ is given by

$$\frac{\mathrm{d}p^\sigma_{RT_u}(r,t,0)}{\mathrm{d}\tau} = \frac{1}{|\Omega|} \int_\Omega \partial_2 K_\sigma(r - R(x), t - T_u(x)) \, \nabla T_u(x) \cdot v(x) \, \mathrm{d}x.$$

Defining $E^\sigma_{RT_u}(r,t) := 1 - \log \frac{p^\sigma_{RT_u}(r,t)}{p^\sigma_R(r) \, p^\sigma_{T_u}(t)}$ we obtain

$$\left. \frac{\mathrm{d}\mathcal{D}^{\mathrm{MI}_\sigma}[u + \tau v]}{\mathrm{d}\tau} \right|_{\tau=0}$$

$$= \int_{\mathbb{R}^2} \left(\frac{1}{|\Omega|} \int_\Omega \partial_2 K_\sigma(r - R(x), t - T_u(x)) \, \nabla T_u(x) \cdot v(x) \, \mathrm{d}x \right) L(r,t) \, \mathrm{d}(r,t)$$

$$= \frac{1}{|\Omega|} \int_\Omega \left(\int_{\mathbb{R}^2} \partial_2 K_\sigma(r - R(x), t - T_u(x)) \, E^\sigma_{RT_u}(r,t) \, \mathrm{d}(r,t) \right) \nabla T_u(x) \cdot v(x) \, \mathrm{d}x$$

$$= \frac{1}{|\Omega|} \int_\Omega \left(\partial_2 K_\sigma * E^\sigma_{RT_u} \right) (R(x), T_u(x)) \, \nabla T_u(x) \cdot v(x) \, \mathrm{d}x$$

and integrating by parts results $\partial_2 K_\sigma * E^\sigma_{RT_u} = -K_\sigma * \partial_2 E^\sigma_{RT_u}$ with

$$\partial_2 E^\sigma_{RT_u}(r,t) = \frac{p^{\sigma\prime}_{T_u}(t)}{p^\sigma_{T_u}(t)} - \frac{\partial_2 p^\sigma_{RT_u}(r,t)}{p^\sigma_{RT_u}(r,t)}.$$

Defining $L^\sigma_{RT_u} := -\frac{1}{|\Omega|} \partial_2 E^\sigma_{RT_u}$ completes the proof.   ∎

Note that the force of the approximating approach does not coincide with the approximation (4.6) of the forces from the direct approach. An additional convolution with the kernel of the approximating densities occurs. This is a direct consequence of first approximating the MI functional and then computing the derivative in a second step. In contrast, in the direct approach we first compute the derivative and subsequently form approximations. However, both approaches lead to quite similar expressions, whereas the direct approach allows for a wide class of approximating schemes.

Now we compute the Gateâux derivatives for the smoothers and establish the classical formulation of the Euler-Lagrange equations for the registration problem as boundary value problem. We start with the diffusive smoother.

4.2.2 Diffusive Registration

The diffusive smoother was defined as (cf. (1.4) on page 10)

$$\mathcal{S}^{\mathrm{DIFF}}[u] := \frac{1}{2} \sum_{\ell=1}^d \int_\Omega \|\nabla u_\ell\|^2 \, \mathrm{d}x = \frac{1}{2} \langle \nabla u, \nabla u \rangle_{L^2(\Omega)} \qquad (4.7)$$

where ∇u is a shorthand notation for the component-wise application of ∇ to $u : \mathbb{R}^d \to \mathbb{R}^d$ in the manner $\nabla u = (\nabla u_1, \nabla u_2, \dots, \nabla u_d)^\top : \mathbb{R}^d \to \mathbb{R}^{2d}$.

Lemma 4.4 (Gateâux Derivative of the Diffusive Smoother)
Let $u, v \in C^2(\overline{\Omega}; \mathbb{R}^d)$. The Gateâux derivative of $\mathcal{S}^{\mathrm{DIFF}}$ is given by

$$d\mathcal{S}^{\mathrm{DIFF}}[u; v] = \langle \mathscr{A}u, v \rangle_{L^2(\Omega)} + \sum_{\ell=1}^{d} \int_{\partial\Omega} v_\ell \, \langle \nabla u_\ell, n \rangle \, \mathrm{d}S$$

with $\mathscr{A} = \mathscr{A}^{\mathrm{DIFF}} := -\Delta$ the negative Laplace operator.

Proof. Let $u, v \in C^1(\overline{\Omega}; \mathbb{R}^d)$. A direct computation shows

$$\mathcal{S}^{\mathrm{DIFF}}[u + \tau v] =$$

$$\frac{1}{2} \sum_{\ell=1}^{d} \int_{\Omega} \|\nabla u_\ell\|^2 \, \mathrm{d}x + \tau \sum_{\ell=1}^{d} \int_{\Omega} \nabla u_\ell \, \nabla v_\ell \, \mathrm{d}x + \frac{\tau^2}{2} \sum_{\ell=1}^{d} \int_{\Omega} \|\nabla v_\ell\|^2 \, \mathrm{d}x.$$

Differentiating w.r.t. τ and setting $\tau = 0$ results

$$d\mathcal{S}^{\mathrm{DIFF}}[u; v] = \sum_{\ell=1}^{d} \int_{\Omega} \langle \nabla u_\ell, \nabla v_\ell \rangle \, \mathrm{d}x.$$

For $u, v \in C^2(\overline{\Omega}; \mathbb{R}^d)$ we can apply Green's formula such that

$$\int_{\Omega} \langle \nabla u_\ell, \nabla v_\ell \rangle \, \mathrm{d}x = \int_{\partial\Omega} v_\ell \, \langle \nabla u_\ell, n \rangle \, \mathrm{d}S - \int_{\Omega} v_\ell \, \Delta u_\ell \, v_\ell \, \mathrm{d}x.$$

∎

Once having computed the Gateâux derivatives of the smoother and a distance measure we can directly read off the boundary value problem for the registration problem.

Let $\mathcal{D}$ be one of the two distance measures above with $d\mathcal{D}[u; v] = \langle f(u), v \rangle_{L^2(\Omega)}$ and $\mathcal{J} := \mathcal{D} + \alpha \mathcal{S}^{\mathrm{DIFF}}$ the joint functional for the diffusive registration. Then the Gateâux derivative of $\mathcal{J}$ is given by

$$d\mathcal{J}[u; v] = d\mathcal{D}[u; v] + \alpha \, d\mathcal{S}^{\mathrm{DIFF}}[u; v]$$

$$= \langle f(u) + \alpha \mathscr{A}u, v \rangle_{L^2(\Omega)} + \alpha \sum_{\ell=1}^{d} \int_{\partial\Omega} v_\ell \, \langle \nabla u_\ell, n \rangle \, \mathrm{d}S.$$

The Euler-Lagrange equations for minimizers $u \in C^2(\overline{\Omega}; \mathbb{R}^d)$ of $\mathcal{J}$ reads $d\mathcal{J}[u; v] = d\mathcal{D}[u; v] + \alpha \, d\mathcal{S}^{\mathrm{DIFF}}[u; v] = 0$ for all $v \in C^2(\overline{\Omega}; \mathbb{R}^d)$. In particular they must hold for all $v \in C_c^\infty(\Omega; \mathbb{R}^d) \subset C^2(\overline{\Omega}; \mathbb{R}^d)$. Since $v = 0$ on $\partial\Omega$ for $v \in C_c^\infty(\Omega; \mathbb{R}^d)$ the boundary integrals vanish and we find

$$d\mathcal{J}[u; v] = \langle f(u) + \alpha \mathscr{A}u, v \rangle_{L^2(\Omega)} = 0 \text{ for all } v \in C_c^\infty(\Omega)$$

$$\overset{\text{Lemma 4.1}}{\Leftrightarrow} \quad f(u) + \alpha \mathscr{A}u = 0 \text{ on } \Omega.$$

Furthermore, $d\mathcal{J}[u; v] = 0$ must hold for all $v \in C^\infty(\overline{\Omega}; \mathbb{R}^d) \subset C^2(\overline{\Omega}; \mathbb{R}^d)$ with $v_k = 0$ on $\partial\Omega$ for all $k \in \{1, 2, \ldots, d\} \setminus \{\ell\}$ such that almost every boundary integral vanishes. Since we already know $f(u) + \alpha \mathscr{A} u = 0$ we find

$$d\mathcal{J}[u; v] = \alpha \int_{\partial\Omega} v_\ell \langle \nabla u_\ell, n \rangle \, dS = 0 \text{ for all } v_\ell \in C^\infty(\overline{\Omega}; \mathbb{R}^d)$$

$$\overset{\text{Lemma 4.1}}{\Leftrightarrow} \quad \langle \nabla u_\ell, n \rangle = 0 \text{ on } \partial\Omega.$$

Thus, a minimizer must fulfill the so-called *Neumann boundary conditions* $\langle \nabla u_\ell, n \rangle = 0$ on $\partial\Omega$ for $\ell = 1, 2, \ldots, d$.
Concluding, a (local) minimizer of $\mathcal{J}$ on $C^2(\overline{\Omega}; \mathbb{R}^d)$ is a solution of the boundary value problem

$$f(u) + \mathscr{A} u = 0 \qquad \text{on } \Omega$$
$$\text{and} \quad \langle \nabla u_\ell, n \rangle = 0 \qquad \text{on } \partial\Omega, \ \ell = 1, 2, \ldots, d.$$

The boundary conditions are called *natural boundary conditions*, since any minimizer must fulfill them. Alternatively, we can impose *explicit boundary conditions* on the admissible functions and therefore restricting our search space for minimizers. In order to make the boundary integrals vanish we define the search space

$$V_0 := \{u \in C^2(\overline{\Omega}; \mathbb{R}^d) : u = 0 \text{ on } \partial\Omega\} \subset C^2(\overline{\Omega}; \mathbb{R}^d).$$

The imposed explicit boundary condition $u = 0$ on $\partial\Omega$ is called *Dirichlet boundary condition*. For $u, v \in V_0$ we have $d\mathcal{J}[u; v] = \langle f(u) + \alpha \mathscr{A} u, v \rangle_{L^2(\Omega)}$ and therefore $d\mathcal{J}[u; v] = 0$ holds for all $v \in V_0$ if and only if $f(u) + \alpha \mathscr{A} u = 0$. Hence, a (local) minimizer of $\mathcal{J}$ on V_0 is a solution of

$$f(u) + \mathscr{A} u = 0 \quad \text{on } \Omega.$$

Summarizing, we obtain:

Theorem 4.5 (BVP for Diffusive Registration)
Let $\mathcal{D}$ be a distance measure with force f and $\mathcal{J} := \mathcal{D} + \alpha \mathcal{S}^{\text{DIFF}}$.

a) A minimizer $u \in \{v \in C^2(\overline{\Omega}; \mathbb{R}^d) : v = 0 \text{ on } \partial\Omega\}$ of $\mathcal{J}$ is a solution of

$$f(u) - \alpha \Delta u = 0 \quad \text{on } \Omega. \qquad\qquad \textit{(Dirichlet problem)}$$

b) A minimizer $u \in C^2(\overline{\Omega}; \mathbb{R}^d)$ of $\mathcal{J}$ is a solution of

$$f(u) - \alpha \Delta u = 0 \quad \text{on } \Omega$$

and $\langle \nabla u_\ell, n \rangle = 0$ on $\partial\Omega$ for $\ell = 1, 2, \ldots, d.$ $\qquad$ *(Neumann problem)*

Remark 4.6 (Neumann Boundary Conditions for $\Omega = (0,1)^d$)

For $\Omega = (0,1)^d$ the outer normal field $n(x)$ is always a unit vector $e_j = (\delta_{ij})_{i=1}^d \in \mathbb{R}^d$ with

$$\delta_{ij} = \begin{cases} 1 & \text{if } i = j, \\ 0 & \text{otherwise.} \end{cases}$$

Therefore, for the Neumann boundary condition holds

$$\langle \nabla u_\ell, n \rangle = 0 \text{ on } \partial(0,1)^d \quad \Leftrightarrow \quad \partial_j u_\ell(x) = 0 \text{ if } n(x) = e_j, \text{ for all } x \in \partial(0,1)^d.$$

4.2.3 Curvature Registration

The curvature smoother was defined as (cf. (1.5) on page 11)

$$\mathcal{S}^{\text{CURV}}[u] := \frac{1}{2} \sum_{\ell=1}^d \int_\Omega (\Delta u_\ell)^2 \, \mathrm{d}x = \frac{1}{2} \langle \Delta u, \Delta u \rangle_{L^2(\Omega)}. \tag{4.8}$$

where Δu denotes the application of the Laplace operator to each component of $u : \mathbb{R}^d \to \mathbb{R}^d$, that is $\Delta u = (\Delta u_1, \Delta u_2, \ldots, \Delta u_d)^\top : \mathbb{R}^d \to \mathbb{R}^d$.

Lemma 4.7 (Gateâux Derivative of the Curvature Smoother)
Let $u, v \in C^2(\overline{\Omega}; \mathbb{R}^d)$. The Gateâux Derivative of $\mathcal{S}^{\text{CURV}}$ is given by

$$d\mathcal{S}^{\text{CURV}}[u;v] = \langle \mathscr{A} u, v \rangle_{L^2(\Omega)}$$

$$+ \sum_{\ell=1}^d \int_{\partial\Omega} \Delta u_\ell \, \langle \nabla v_\ell, n \rangle \, \mathrm{d}S - \sum_{\ell=1}^d \int_{\partial\Omega} v_\ell \, \langle \nabla \Delta u_\ell, n \rangle \, \mathrm{d}S.$$

with $\mathscr{A} = \mathscr{A}^{\text{CURV}} := \Delta^2$ the biharmonic operator.

Proof. Let $u, v \in C^2(\overline{\Omega}; \mathbb{R}^d)$. A direct computation shows

$$\mathcal{S}^{\text{CURV}}[u + \tau v] =$$

$$\frac{1}{2} \sum_{\ell=1}^d \int_\Omega (\Delta u_\ell)^2 \, \mathrm{d}x + \tau \sum_{\ell=1}^d \int_\Omega \Delta u_\ell \, \Delta v_\ell \, \mathrm{d}x + \frac{\tau^2}{2} \sum_{\ell=1}^d \int_\Omega (\Delta v_\ell)^2 \, \mathrm{d}x.$$

Differentiating w.r.t. τ and setting $\tau = 0$ results

$$d\mathcal{S}[u;v]^{\text{CURV}} = \sum_{\ell=1}^d \int_\Omega \Delta u_\ell \, \Delta v_\ell \, \mathrm{d}x.$$

For $u, v \in C^4(\overline{\Omega}; \mathbb{R}^d)$ we can apply Green's formula twice such that

$$\int_\Omega \Delta u_\ell \, \Delta v_\ell \, \mathrm{d}x = \int_{\partial\Omega} \Delta u_\ell \, \langle \nabla v_\ell, n \rangle \, \mathrm{d}S - \int_{\partial\Omega} v_\ell \, \langle \nabla \Delta u_\ell, n \rangle \, \mathrm{d}S + \int_\Omega \Delta^2 u_\ell \, v_\ell \, \mathrm{d}x.$$

■

As for the diffusive registration let $\mathcal{D}$ be a distance measure with force f such that the Gateâux differential for $u, v \in C^0(\Omega; \mathbb{R}^d)$ reads $d\mathcal{D}[u; v] = \langle f(u), v \rangle_{L^2(\Omega)}$. Then the joint functional for the curvature registration is given by $\mathcal{J} := \mathcal{D} + \alpha \mathcal{S}^{\mathrm{CURV}}$ with Gateâux derivative

$$
\begin{aligned}
d\mathcal{J}[u; v] &= d\mathcal{D}[u; v] + \alpha \mathcal{S}^{\mathrm{CURV}}[u; v] \\
&= \langle f(u) + \mathscr{A}u, v \rangle_{L^2(\Omega)} \\
&\quad + \sum_{\ell=1}^{d} \int_{\partial\Omega} \Delta u_\ell \, \langle \nabla v_\ell, n \rangle \, \mathrm{d}S - \sum_{\ell=1}^{d} \int_{\partial\Omega} v_\ell \, \langle \nabla \Delta u_\ell, n \rangle \, \mathrm{d}S.
\end{aligned}
$$

If $u \in C^4(\overline{\Omega}; \mathbb{R}^d)$ is a minimizer of $\mathcal{J}$ then $d\mathcal{J}[u; v] = 0$ must be valid for all $v \in C^4(\overline{\Omega}; \mathbb{R}^d)$ and in particular for $v \in C_c^\infty(\overline{\Omega}; \mathbb{R}^d)$. Then the boundary integrals vanish and we find

$$
d\mathcal{J}[u; v] = 0 \quad \Leftrightarrow \quad f(u) + \alpha \mathscr{A}u = 0 \text{ on } \Omega.
$$

Moreover, choosing $v \in C^\infty(\Omega; \mathbb{R}^d) \subset C^4(\overline{\Omega}; \mathbb{R}^d)$ with both $v_k = 0$, $k = 1, 2, \ldots, d$ and $\langle \nabla v_k, n \rangle = 0$, $k \neq \ell$ on $\partial\Omega$ implies in combination with $f(u) + \alpha \mathscr{A}u = 0$

$$
d\mathcal{J}[u; v] = 0 \quad \Leftrightarrow \quad \Delta u_\ell = 0 \text{ on } \partial\Omega.
$$

Finally, $d\mathcal{J}[u; v] = 0$ must also hold for all $v \in C^\infty(\Omega; \mathbb{R}^d) \subset C^4(\overline{\Omega}; \mathbb{R}^d)$ with both $\langle \nabla v_k, n \rangle = 0$, $k = 1, 2, \ldots, d$ and $v_k = 0$, $k \neq \ell$ on $\partial\Omega$ and therefore

$$
d\mathcal{J}[u; v] = 0 \quad \Leftrightarrow \quad \langle \nabla \Delta u_\ell, n \rangle = 0 \text{ on } \partial\Omega.
$$

Thus, a minimizer $u \in C^4(\overline{\Omega}; \mathbb{R}^d)$ if $\mathcal{J}$ fulfills

$$
f(u) + \mathscr{A}u = 0 \quad \text{on } \Omega
$$

and the natural boundary conditions

$$
\Delta u_\ell = \langle \nabla \Delta u_\ell, n \rangle = 0 \quad \text{on } \partial\Omega, \quad \ell = 1, 2, \ldots, d.
$$

For numerical purposes we take advantage from replacing a natural boundary by an explicit boundary condition. To this end we define the search spaces

$$
\begin{aligned}
V_0 &:= \{v \in C^4(\overline{\Omega}; \mathbb{R}^d) \; : \; v_\ell = 0 \text{ on } \partial\Omega \text{ for } \ell = 1, 2, \ldots, d\}, \\
V_1 &:= \{v \in C^4(\overline{\Omega}; \mathbb{R}^d) \; : \; \langle \nabla v_\ell, n \rangle = 0 \text{ on } \partial\Omega \text{ for } \ell = 1, 2, \ldots, d\}.
\end{aligned} \tag{4.9}
$$

Then, for $u, v \in V_0$ the Gateâux derivative reads

$$
d\mathcal{J}[u; v] = \langle f(u) + \mathscr{A}u, v \rangle_{L^2(\Omega)} + \sum_{\ell=1}^{d} \int_{\partial\Omega} \Delta u_\ell \, \langle \nabla v_\ell, n \rangle \, \mathrm{d}S
$$

and a minimizer $u \in V_0$ must be a solution of $f(u) + \mathscr{A}u = 0$ on Ω with natural boundary conditions

$$\Delta u_\ell = 0 \text{ on } \partial\Omega, \quad \ell = 1, 2, \ldots, d.$$

Thus, we have replaced the natural boundary condition $\langle \nabla \Delta u_\ell, n \rangle = 0$ with the explicit boundary condition $u_\ell = 0$.

Furthermore, choosing the search space V_1, we have

$$d\mathcal{J}[u; v] = \langle f(u) + \mathscr{A}u, v \rangle_{L^2(\Omega)} - \sum_{\ell=1}^{d} \int_{\partial\Omega} v_\ell \, \langle \nabla \Delta u_\ell, n \rangle \, \mathrm{d}S.$$

and a minimizer $u \in V_1$ must be a solution of $f(u) + \mathscr{A}u = 0$ on Ω with natural boundary conditions

$$\langle \nabla \Delta u_\ell, n \rangle = 0 \text{ on } \partial\Omega, \quad \ell = 1, 2, \ldots, d.$$

Hence, the natural boundary condition $\Delta u_\ell = 0$ has been replaced by the explicit boundary condition $\langle \nabla u_\ell, n \rangle = 0$. Summarizing we obtain:

Theorem 4.8 (BVP for Curvature Registration)

Let $\mathcal{D}$ be a distance measure with force f, $\mathcal{J} := \mathcal{D} + \alpha S^{\mathrm{CURV}}$ and V_0, V_1 the function spaces given by (4.9).

a) A minimizer $u \in C^4(\overline{\Omega}; \mathbb{R}^d)$ of $\mathcal{J}$ is a solution of

$$f(u) + \alpha \Delta^2 u = 0 \quad \text{on } \Omega$$

and $\Delta u_\ell = \langle \nabla \Delta u_\ell, n \rangle = 0$ on $\partial\Omega$ for $\ell = 1, 2, \ldots, d$.

b) A minimizer $u \in V_0$ of $\mathcal{J}$ is a solution of

$$f(u) + \alpha \Delta^2 u = 0 \quad \text{on } \Omega$$

and $\Delta u_\ell = 0$ on $\partial\Omega$ for $\ell = 1, 2, \ldots, d$.   (Dirichlet problem)

c) A minimizer $u \in V_1$ of $\mathcal{J}$ is a solution of

$$f(u) + \alpha \Delta^2 u = 0 \quad \text{on } \Omega$$

and $\langle \nabla \Delta u_\ell, n \rangle = 0$ on $\partial\Omega$ for $\ell = 1, 2, \ldots, d$.   (Neumann problem)

Remark 4.9 (Boundary Conditions for $\Omega = (0, 1)^d$)

For $\Omega = (0, 1)^d$ the outer normal field $n(x)$ is always a unit vector e_j (cf. Remark 4.6). Let $u \in V_0$ with $u = 0$ on $\partial\Omega$, $x \in \partial\Omega$ be an arbitrary point with $n(x) = e_j$, and $\gamma(t) := u(x + te_k)$, $k \neq j$. Then $\gamma(t) = 0$ for $-x_k < t < 1 - x_k$ and therefore $\partial_k^m u_\ell(x) = 0$, $m = 1, 2, 3, 4$ for $u \in C^4(\overline{\Omega}; \mathbb{R}^d)$. Thus, for the natural boundary condition $\Delta u_\ell = 0$ on $\partial\Omega$ holds

$$\Delta u_\ell = 0 \text{ on } \partial(0, 1)^d \quad \Leftrightarrow \quad \partial_j^2 u_\ell(x) = 0 \text{ if } n(x) = e_j, \text{ for all } x \in \partial(0, 1)^d.$$

Analogous, let $u \in V_1$ with $\langle \nabla u_\ell, n \rangle = 0$ on $\partial\Omega$ and $x \in \partial\Omega$ be an arbitrary point with $n(x) = e_j$ such that $\nabla u_\ell, n = \partial_j u_\ell$. Defining $\gamma(t) := (\partial_j u)(x + te_k)$, $k \neq j$ we have $\gamma(t) = 0$ for $-x_k < t < 1 - x_k$ and therefore $\partial_k^m (\partial_j u_\ell)(x) = 0$, $m = 1, 2, 3$. Thus, $\langle \nabla \Delta u_\ell(x), n(x) \rangle = \partial_j \Delta u_\ell(x) = \Delta(\partial_j u_\ell)(x)$ and we have the equivalence

$$\langle \nabla \Delta u_\ell, n \rangle = 0 \text{ on } \partial(0,1)^d \quad \Leftrightarrow \quad \partial_j^3 u_\ell(x) = 0 \text{ if } n(x) = e_j, \text{ for all } x \in \partial(0,1)^d.$$

For a compact notation, we introduce the following notation of directional derivatives into the directions of the outer normal field n. Let Ω be a domain and $u \in C^m(\overline{\Omega})$. We define the m-th *normal derivative* of u by

$$\partial_n^m u(x) := \frac{d^m}{d\tau^m} u(x - \tau n(x)) \quad \text{for } x \in \partial\Omega. \tag{4.10}$$

As usual, we also write $\partial_n^2 = \partial_{nn}$, $\partial_n^3 = \partial_{nnn}$, etc.
In particular, for $\Omega = (0,1)^d$ the unit square we have $\partial_n^m u = \partial_j^m u$ if $n = e_j$. Thus, Remark 4.6 and Remark 4.9 read

$$\langle \nabla u, n \rangle = 0 \text{ on } \partial(0,1)^d \quad \Leftrightarrow \quad \partial_n u = 0 \text{ on } \partial(0,1)^d$$
$$\Delta u = 0 \text{ on } \partial(0,1)^d \quad \Leftrightarrow \quad \partial_{nn} u = 0 \text{ on } \partial(0,1)^d,$$
$$\langle \nabla \Delta u, n \rangle = 0 \text{ on } \partial(0,1)^d \quad \Leftrightarrow \quad \partial_{nnn} u = 0 \text{ on } \partial(0,1)^d.$$

The idea of the curvature smoother is that affine transformation should not be penalized. Nevertheless, the functions in the search space V_0 for the Dirichlet BVP and the first order normal derivatives of functions in the search space V_1 for the Neumann BVP must vanish on the boundary of Ω. Thus, V_0 includes only a single affine transformation, namely $v(x) = 0$, and V_1 covers only affine transformation that are constant, i.e., $v(x) = c$. Therefore, we can only expect that solutions of the BVPs are smooth but in general not to be an affine transformation.
To overcome this problem, in [27] the author proposes the "full curvature" approach using

$$S[u] = \sum_{\ell=1}^{d} \int_\Omega \sum_{i,j=1}^{d} (\partial_{ij} u_\ell)^2 \ dx$$

as smoother. It turns out that classical solutions of the full curvature registration must satisfy the biharmonic equation $f(u) + \alpha \Delta^2 u = 0$ on Ω, too, but the associated boundary conditions involve only second order homogeneous derivatives. Thus, they are naturally fulfilled by any affine transformation. Nevertheless, these boundary conditions are much more complicated than ours and their numerical treatment much more involved.

4.2.4 Elastic Registration

The elastic smoother was given by (cf. (1.3) on page 10)

$$\mathcal{S}^{\mathrm{ELAS}}[u] = \frac{1}{2}\int_{\Omega}\sum_{j,k=1}^{d}\frac{\mu}{2}(\partial_j u_k + \partial_k u_j)^2 + \lambda(\mathrm{div}\,u)^2\,\mathrm{d}x = \frac{1}{2}\langle\mathscr{L}u,\mathscr{L}u\rangle_{L^2(\Omega)}\ (4.11)$$

where $\mathscr{L}u$ is defined for $u \in C^1(\Omega;\mathbb{R}^d)$ as

$$\mathscr{L}u = \frac{1}{\sqrt{2}}\begin{pmatrix}\sqrt{\mu}\,(\nabla u_1 + \partial_1 u)\\ \sqrt{\mu}\,(\nabla u_2 + \partial_2 u)\\ \vdots\\ \sqrt{\mu}\,(\nabla u_d + \partial_d u)\\ \sqrt{2\lambda}\,\mathrm{div}\,u\end{pmatrix} : \mathbb{R}^d \to \mathbb{R}^{2d+1}$$

with $\partial_k u = (\partial_k u_1, \partial_k u_2, \ldots, \partial_k u_d)^\top : \mathbb{R}^d \to \mathbb{R}^d$, $k = 1, 2, \ldots, d$.

Lemma 4.10 (Gateâux Derivative of the Elastic Smoother)
Let $u, v \in C^2(\overline{\Omega};\mathbb{R}^d)$. The Gateâux derivative of $\mathcal{S}^{\mathrm{ELAS}}$ is given by

$$d\mathcal{S}^{\mathrm{ELAS}}[u;v] = \langle\mathscr{A}u,v\rangle_{L^2(\Omega)} + \mu\sum_{k=1}^{d}\int_{\partial\Omega}v_k\,\langle\nabla u_k + \partial_k u, n\rangle\,\mathrm{d}S$$

$$+ \lambda\int_{\partial\Omega}\mathrm{div}\,u\,\langle v, n\rangle\,\mathrm{d}S$$

with $\mathscr{A} = \mathscr{A}^{\mathrm{ELAS}}$ the so-called Navier-Lamé operator

$$\mathscr{A}^{\mathrm{ELAS}} := -(\mu\Delta + (\lambda + \mu)\nabla\,\mathrm{div}).$$

Proof. The elastic smoother can be written as $\mathcal{S}^{\mathrm{ELAS}}[u] = \frac{1}{2}a[u,u]$ with

$$a[u,v] = \int_{\Omega}\lambda\,\mathrm{div}\,u\,\,\mathrm{div}\,v + \frac{\mu}{2}\sum_{j,k=1}^{d}(\partial_j u_k + \partial_k u_j)(\partial_j v_k + \partial_k v_j)\,\mathrm{d}x.$$

The functional $a[\cdot,\cdot]$ is symmetric and linear in its arguments, that is $a[u,v] = a[v,u]$, $a[\lambda u,v] = \lambda a[u,v]$ and $a[u+v,w] = a[u,w] + a[v,w]$. Thus,

$$\mathcal{S}^{\mathrm{ELAS}}[u + \tau v] = \tfrac{1}{2}a[u,u] + \tau a[u,v] + \tfrac{\tau^2}{2}a[v,v]$$

and differentiating w.r.t. τ and setting $\tau = 0$ yields

$$d\mathcal{S}^{\mathrm{ELAS}}[u;v] = a[u,v].$$

Moreover, defining $F := \sum_{k=1}^{d}v_k\,(\nabla u_k + \partial_k u)$ and $G := v\,\mathrm{div}\,u$ it is straightforward to verify

$$\mathrm{div}\,F = \frac{1}{2}\sum_{j,k=1}^{d}(\partial_j u_k + \partial_k u_j)(\partial_j v_k + \partial_k v_j) + \langle\Delta u + \nabla\,\mathrm{div}\,u, v\rangle,$$

$$\mathrm{div}\,G = \mathrm{div}\,v\,\,\mathrm{div}\,u + \langle\nabla\,\mathrm{div}\,u, v\rangle.$$

From the divergence theorem we have

$$\int_{\partial\Omega} \langle \mu F + \lambda G, n \rangle \ \mathrm{d}S = \int_{\Omega} \mathrm{div}(\mu F + \lambda G) \ \mathrm{d}x = \int_{\Omega} \mu \, \mathrm{div}\, F + \lambda \, \mathrm{div}\, G \ \mathrm{d}x.$$

Splitting and rearranging the integral on Ω results

$$a[u, v] = -\int_{\Omega} \langle \mu \Delta u + (\lambda + \mu) \nabla \, \mathrm{div}\, u, v \rangle \ \mathrm{d}x$$

$$+ \int_{\partial\Omega} \mu \sum_{k=1}^{d} v_k \, \langle \nabla u_k + \partial_k u, n \rangle + \lambda \, \mathrm{div}\, u \, \langle v, n \rangle \ \mathrm{d}S$$

$$= \langle \mathscr{A} u, v \rangle_{L^2(\Omega)} + \mu \sum_{k=1}^{d} \int_{\partial\Omega} v_k \, \langle \nabla u_k + \partial_k u, n \rangle \ \mathrm{d}S$$

$$+ \lambda \int_{\partial\Omega} \mathrm{div}\, u \, \langle v, n \rangle \ \mathrm{d}S.$$

∎

In line with our previous analysis, let $\mathcal{D}$ be a distance measure with force f and $\mathcal{J} := \mathcal{D} + \alpha \mathcal{S}^{\mathrm{ELAS}}$ the joint functional for the elastic registration.

In the former sections on diffusive and curvature registration we searched minimizers in $C^2(\overline{\Omega}; \mathbb{R}^d)$ and $C^4(\overline{\Omega}; \mathbb{R}^d)$ respectively, and found that a minimizer has to fulfill natural boundary conditions. Later on, we saw that natural boundary conditions can by replaced by explicit boundary conditions imposed on all functions in the search space. In principle, we can proceed for the elastic registration as well and search for minimizers in $C^2(\overline{\Omega}; \mathbb{R}^d)$. Unfortunately, the occuring natural boundary conditions look complicated and are quite impractical for the design of numerical schemes. By this, we skip the analysis for minimizers in $C^2(\overline{\Omega}; \mathbb{R}^d)$ and restrict ourselves to the search space

$$V_0 := \{ v \in C^2(\overline{\Omega}; \mathbb{R}^d) \ : \ v = 0 \quad \text{on } \partial\Omega \}.$$

Thus, we impose the explicit boundary condition $v = 0$ on $\partial\Omega$ and replace the above mentioned natural boundary condition.

However, for $u, v \in V_0$ the Gateâux differential of the joint functional $\mathcal{J}$ reads

$$d\mathcal{J}[u; v] = d\mathcal{D}[u; v] + \alpha d\mathcal{S}^{\mathrm{ELAS}}[u; v] = \langle f(u) + \mathscr{A} u, v \rangle_{L^2(\Omega)}.$$

If u is a minimizer then $d\mathcal{J}[u; v] = 0$ is valid for all $v \in V_0$ and in particular for $v \in C_c^{\infty}(\Omega) \subset V_0$. Hence, Lemma 4.1 implies

$$d\mathcal{J}[u; v] = 0 \text{ for all } v \in V_0 \quad \Leftrightarrow \quad f(u) + \mathscr{A} u = 0 \text{ on } \Omega.$$

Theorem 4.11 (BVP for Elastic Registration)
Let $\mathcal{D}$ be a distance measure with force f and $\mathcal{J} := \mathcal{D} + \alpha \mathcal{S}^{\text{ELAS}}$.
A minimizer $u \in \{v \in C^2(\overline{\Omega}; \mathbb{R}^d) \ : \ v = 0 \text{ on } \partial\Omega\}$ of $\mathcal{J}$ is a solution of

$$f(u) - \alpha(\mu\Delta + (\lambda + \mu)\nabla\operatorname{div})u = 0 \quad \text{on } \Omega. \qquad \textit{(Dirichlet problem)}$$

An interesting property of the Navier-Lamé operator $\mathscr{A}^{\text{ELAS}}$ is that it is not uniquely determined by the elastic energy functional $\mathcal{S}^{\text{ELAS}}$. Let

$$\widehat{\mathcal{S}^{\text{ELAS}}}[u] := \frac{1}{2}\int_\Omega \mu\sum_{\ell=1}^{d}\|\nabla u_\ell\|^2 + (\lambda+\mu)(\operatorname{div}u)^2 \ \mathrm{d}x$$

$$= \mu\mathcal{S}^{\text{DIFF}}[u] + \frac{\lambda+\mu}{2}\int_\Omega(\operatorname{div}u)^2 \ \mathrm{d}x. \qquad (4.12)$$

Lemma 4.12 (Gateâux Derivative of $\widehat{\mathcal{S}^{\text{ELAS}}}$)
Let $u, v \in C^2(\overline{\Omega}; \mathbb{R}^d)$. The Gateâux derivative of $\widehat{\mathcal{S}^{\text{ELAS}}}$ is given by

$$d\widehat{\mathcal{S}^{\text{ELAS}}}[u; v] = \langle \mathscr{A}u, v \rangle_{L^2(\Omega)}$$

$$+ \mu\sum_{\ell=1}^{d}\int_{\partial\Omega}v_\ell \langle\nabla u_\ell, n\rangle \ \mathrm{d}S + (\lambda+\mu)\int_{\partial\Omega}\operatorname{div}u \langle v, n\rangle \ \mathrm{d}S$$

with $\mathscr{A} = \mathscr{A}^{\text{ELAS}} = -(\mu\Delta + (\lambda+\mu)\nabla\operatorname{div})$ the Navier-Lamé operator.

Proof. Defining $a[u, v] := \int_\Omega \operatorname{div}u \ \operatorname{div}v \ \mathrm{d}x$ we have

$$\widehat{\mathcal{S}^{\text{ELAS}}}[u] = \mu\mathcal{S}^{\text{DIFF}}[u] + \frac{\lambda+\mu}{2}a[u, u].$$

Thus, the Gateâux derivative of $\widehat{\mathcal{S}^{\text{ELAS}}}$ is given by

$$d\widehat{\mathcal{S}^{\text{ELAS}}}[u; v] = \mu \, d\mathcal{S}[u; v] + (\lambda+\mu)\, a[u, v]$$

and from the poof for the Gateâux derivative of the elastic smoother we have $a[u, v] = \int_{\partial\Omega}\operatorname{div}u \langle v, n\rangle \ \mathrm{d}S - \int_\Omega \langle \nabla \operatorname{div}u, v\rangle \ \mathrm{d}x$. ∎

Looking for minimizers of $\mathcal{D}+\alpha\widehat{\mathcal{S}^{\text{ELAS}}}$ in $C^2(\overline{\Omega}; \mathbb{R}^d)$ leads to different natural boundary conditions as those mentioned above, but they are impractical for (our) numerical purposes, too. However, in literature the registration based on $\widehat{\mathcal{S}^{\text{ELAS}}}$ is also referred to as elastic registration, even if there is no physical motivation from the theory of elasticity behind it. Nevertheless, the above analysis carried out for minimizers $u \in V_0$ of $\mathcal{D} + \alpha\mathcal{S}^{\text{ELAS}}$ is also valid for $\mathcal{D} + \alpha\widehat{\mathcal{S}^{\text{ELAS}}}$.

Theorem 4.13 (BVP for Elastic Registration with $S^{\widehat{ELAS}}$)
Let $\mathcal{D}$ be a distance measure with force f and $\mathcal{J} := \mathcal{D} + \alpha S^{\widehat{ELAS}}$.
A minimizer $u \in \{v \in C^2(\overline{\Omega}; \mathbb{R}^d) \; : \; v = 0 \text{ on } \partial\Omega\}$ of $\mathcal{J}$ is a solution of

$$f(u) - \alpha(\mu\Delta + (\lambda + \mu)\nabla \operatorname{div})u = 0 \quad \text{on } \Omega. \qquad \textit{(Dirichlet problem)}$$

4.3 Gradient-Flow

In this section we investigate the design of a general method to compute a solution of the Euler-Lagrange equations and associated boundary value problems respectively. Table 4.1 gives an overview of the boundary value problems to be solved for the diffusive, curvature, and elastic registration. The derived equations are non-linear, so we cannot expect to compute a solution directly. To this end, we apply a continuous gradient decent strategy.

Definition 4.14 (Gradient)
Let $(X, \langle \cdot, \cdot \rangle_X)$ be a Hilbert space and $X_0 \subseteq X$ a linear subspace. We call $\nabla F(u) \in X$ gradient of $F : X_0 \to \mathbb{R}$ at $u \in X_0$ if

$$\left. \frac{dF[u + \tau v]}{d\tau} \right|_{\tau=0} = dF[u; v] = \langle \nabla F(u), v \rangle_X \qquad \textit{for all } v \in X_0.$$

For image registration we are searching for minimizers of a joint functional $\mathcal{J} = \mathcal{D} + \alpha S$ in a linear space

$$V = \{u \in C^m(\overline{\Omega}) \; : \; u \text{ fulfills natural and/or}$$
$$\text{explicit boundary conditions}\} \subseteq C^m(\overline{\Omega}) \subset L^2(\Omega).$$

The Gateâux derivative of $\mathcal{J}$ is given by

$$d\mathcal{J}[u; v] = \langle f(u) + \alpha \mathcal{A}u, v \rangle_{L^2(\Omega)} \qquad \text{for all } u, v \in V.$$

Thus, the gradient of $\mathcal{J}$ w.r.t. the L^2 inner product is given by

$$\nabla\mathcal{J}(u) = f(u) + \alpha \mathcal{A}u$$

and the Euler-Lagrange equations reveal as the common first order necessary condition for minimizers u of $\mathcal{J}$

$$\nabla\mathcal{J}(u) = 0.$$

From the theory of optimization it is well-known that the negative gradient of a functional points into the direction for which the function value decreases the most. For the design of a continuous gradient descent we

Registration	PDE on $(0,1)^d$	Boundary conditions on $\partial(0,1)^d$	
Diffusive Registration	$f(u) - \alpha \Delta u = 0$	$u = 0$	(Dirichlet)
	$f(u) - \alpha \Delta u = 0$	$\partial_\nu u = 0$	(Neumann)
Curvature Registration	$f(u) + \alpha \Delta^2 u = 0$	$u = \partial_{\nu\nu} u = 0$	(Dirichlet)
	$f(u) + \alpha \Delta^2 u = 0$	$\partial_\nu u = \partial_{\nu\nu\nu} u = 0$	(Neumann)
Elastic Registration	$f(u) - \alpha(\mu\Delta + (\mu + \lambda)\nabla \operatorname{div})u = 0$	$u = 0$	(Dirichlet)

Table 4.1: Boundary value problems for diffusive, curvature, and elastic registration

create a time dependent process. To this end, we initially choose a function u_0 as a starting point and evolve it over time along the direction of the negative gradient of $\mathcal{J}$. More precisely, we make $u : \overline{\Omega} \to \mathbb{R}^d$ time-dependent. Therefore, we introduce an artificial time and additionally let u depend on a time parameter. Thus, we consider

$$u : \overline{\Omega} \times \mathbb{R}_+ \to \mathbb{R}^d, \quad (x,t) \mapsto u(x,t) \quad \text{with} \quad u(x,0) = u_0(x) \text{ for } x \in \overline{\Omega}.$$

For the gradient descent we require $u(\cdot,t)$ evolves at time t into the direction of the negative gradient $-\nabla \mathcal{J}(u(\cdot,t))$, that is

$$\partial_t u = -\nabla \mathcal{J}(u) \quad \text{and} \quad u(\cdot,0) = u_0.$$

When the evolution converges to a steady state u^*, that is $u \to u^*$ with $\partial_t u^* = 0$ as $t \to \infty$, then u^* is a zero of the gradient and hence a solution of the Euler-Lagrange equations.

Summarizing, we compute a solution of the Euler-Lagrange equations for the image registration problem by computing a solution of the initial value problem

$$\partial_t u = -(f(u) + \alpha \mathscr{A} u) \text{ on } \Omega \quad \text{with} \quad u(x,0) = u_0(x) \text{ for } x \in \overline{\Omega}. \quad (4.13)$$

In a physical manner we can interpret u as a system that evolves into a state of equilibrium over time. Therefore, $u(x,t)$ describes the state of the system at the point x and time t, $f(u)$ is a force acting on the system, and $-\alpha \mathscr{A} u$ describes the spatial reaction of the system. Note that calling f force stems from this interpretation. If the system reaches a steady state, then $f(u) = -\alpha \mathscr{A} u$ and the force is balanced with the reaction of the system.

However, to compute a solution of (4.13) we use an Euler method. To this end, we introduce discrete time steps $t_{n+1} = t_n + \tau_n$, $t_0 = 0$ and approximate the partial time derivative by a backward finite difference scheme. Defining $u^n := u(\cdot, t_n)$ we end up with the full implicit scheme

$$\frac{u^{n+1} - u^n}{\tau_n} = -(f(u^{n+1}) + \alpha \mathscr{A} u^{n+1}). \quad (4.14)$$

In order to compute u^{n+1} from u^n by an Euler step, we have to solve a non-linear equation since the non-linear force still depends on u^{n+1}. In general, we cannot solve the non-linearity directly. This was still the motivation for the gradient descent method. To overcome this difficulty, we linearize (4.14) by approximating $f(u^{n+1})$ with the force $f(u^n)$ of the prior step. This results the semi-implicit Euler method

$$(\mathrm{id} + \tau_n \alpha \mathscr{A}) u^{n+1} = u^n - \tau_n f(u^n). \quad (4.15)$$

Recall that (4.15) stems from a continuous gradient descent strategy. Comparing the Euler method with a discrete gradient descent method, clearly the τ_n corresponds to step length.

Moreover, equation (4.15) can be interpreted as the Euler-Lagrange equation of a quadratic approximation of the joint functional $\mathcal{J} = \mathcal{D} + \alpha \mathcal{S}$ with an additional so-called *iterated Tikhonov regularization*. Therefore, we approximate the distance measure $\mathcal{D}$ with its first-order Taylor expansion at u^n,

$$\mathcal{D}[u] \approx \mathcal{D}[u^n] + \langle \nabla \mathcal{D}(u^n), u - u^n \rangle_{L^2(\Omega)} = \mathcal{D}[u^n] + \langle f(u^n), u - u^n \rangle_{L^2(\Omega)}$$

such that $\mathcal{J}[u] \approx \mathcal{D}[u^n] + \langle f(u^n), u - u^n \rangle_{L^2(\Omega)} + \alpha \mathcal{S}[u]$ provides a good approximation if u is in a neighborhood of u^n. To this end, in addition we regularize the problem by adding a distance term between u and u^n. Defining

$$\mathcal{J}_n[u] := \mathcal{D}[u^n] + \langle f(u^n), u - u^n \rangle_{L^2(\Omega)} + \alpha \mathcal{S}[u] + \frac{1}{2\tau} \| u - u^n \|^2_{L^2(\Omega)}$$

we find

$$\nabla \mathcal{J}_n(u) = f(u^n) + \alpha \mathscr{A} u + \frac{1}{\tau}(u - u^n)$$

where τ acts as a regularization parameter. Thus, the Euler-Lagrange equations for a minimizer of $\mathcal{J}_n$ reads

$$(\mathrm{id} + \tau \alpha \mathscr{A})u = u^n - \tau f(u^n).$$

A detailed discussion on the relation of variational approaches and iterated Tikhonov regularization for the case of diffusion filtering is given in [50].

In the following section we discretize the Euler-Lagrange equations and initial value problem respectively, and compute numerical solutions.

4.4 Discretization

For the computation of numerical solutions of the above PDEs we apply the finite difference method. Therefore, we discretize the domain $\Omega \subset \mathbb{R}^d$ by an equispaced grid and approximate derivatives by difference schemes. In particular, we always consider $\Omega = (0,1)^d$ to be the unit cube.

4.4.1 Discretization of Ω

We discretize $\Omega \subset \mathbb{R}^d$ by an equispaced grid and collect the grid-spacings for each dimension in a vector $h = (h_1, h_2, \ldots, h_d)^\top$, where $h_\ell > 0$ is the spacing in the ℓ-th dimension. We define the grid-points

$$x_k := a + h \odot k \quad \text{for } k \in \mathbb{Z}^d, \, h \in \mathbb{R}^d_+, \text{ and an offset } a \in \mathbb{R}^d,$$

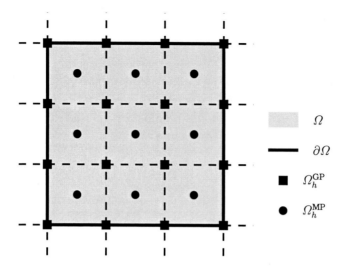

Figure 4.1: Grid and mid-point symmetric discretization ($N_1 = N_2 = 3$)

with the so-called *Hadamard product*

$$x \odot y := (x_1 y_1, x_2 y_2, \ldots, x_d y_d)^\top, \quad \text{for } x, y \in \mathbb{R}^d. \tag{4.16}$$

Furthermore, we define the grid

$$\Omega_h := \{x_k \ : \ k \in \mathbb{Z}^d\} \cap \overline{\Omega}.$$

Moreover, for a short notation, we also index the values of a function $u : \mathbb{R}^d \to \mathbb{R}^d$ at grid points, that is

$$u_k := u(x_k).$$

For the discretization of $\Omega = (0,1)^d$ we choose in particular grid-spacings $h = (1/N_1, 1/N_2, \ldots, 1/N_d)$ with $N_\ell \in \mathbb{N}$, $\ell = 1, 2, \ldots, d$. In the following we consider two particular kinds of grids. The first kind is called *grid-point symmetric*, given by

$$\Omega_h^{GP} := \{x_k = k \odot h \ : \ k_\ell = 0, 1, \ldots, N_\ell, \ \ell = 1, 2, \ldots, d\},$$

and the second kind is called *mid-point symmetric*, given by

$$\Omega_h^{MP} := \{x_k = h/2 + k \odot h \ : \ k_\ell = 0, 1, \ldots, N_\ell - 1, \ \ell = 1, 2, \ldots, d\}.$$

Figure 4.1 shows an example for the discretization of $\Omega = (0,1)^2$ with $h_1 = h_2 = 1/3$.

The grid-symmetry does not play a crucial role in the following. Generally, we can use both discretizations for the design of numerical methods based on finite differences. Nevertheless, the symmetry must be taken into account in the modeling of discrete analogons of boundary conditions as well as for the interpolation used for multilevel and multigrid methods.

However, the choice of a particular discretization depends on the application, the modeling, and implementation aspects. We will not discuss this any further.

4.4.2   Finite Differences

For the approximation of the differential operators $\mathscr{A}^{\mathrm{DIFF}}$, $\mathscr{A}^{\mathrm{CURV}}$, and $\mathscr{A}^{\mathrm{ELAS}}$ we use finite differences. Recall that these operators are defined as

$$\mathscr{A}^{\mathrm{DIFF}} = -\Delta = -\sum_{j=1}^{d} \partial_{jj}, \quad \mathscr{A}^{\mathrm{CURV}} = \Delta^2 = \sum_{j=1}^{d}\sum_{k=1}^{d} \partial_{jj}\partial_{kk}, \tag{4.17}$$

with the implicit understanding of applying the operators componentwise, and $\mathscr{A}^{\mathrm{ELAS}} = \mu\Delta + (\lambda+\mu)\nabla\,\mathrm{div}$ which, for example, in the two dimensional case reads

$$\mathscr{A}^{\mathrm{ELAS}} = -\begin{pmatrix} (2\mu+\lambda)\partial_{11} + \mu\partial_{22} & (\lambda+\mu)\partial_{12} \\ (\lambda+\mu)\partial_{21} & \mu\partial_{11} + (2\mu+\lambda)\partial_{22} \end{pmatrix}. \tag{4.18}$$

The approximations we will derive are based on the basic difference operators ∂_j^{h+}, ∂_j^{h-}, and $\partial_j^{h\pm}$ defined as

$$\partial_j^{h+} u(x) := \frac{u(x+h_j e_j) - u(x)}{h_j}, \qquad \text{(forward difference)}$$

$$\partial_j^{h-} u(x) := \frac{u(x) - u(x-h_j e_j)}{h_j}, \qquad \text{(backward difference)}$$

$$\partial_j^{h\pm} u(x) := \frac{u(x+h_j e_j) - u(x-h_j e_j)}{2h_j}. \qquad \text{(central difference)}$$

Expanding a three-times continuous differentiable function u at x in its Taylor series $u(x \pm h_j e_j) = u(x) \pm h_j \partial_j u(x) + \frac{h_j^2}{2}\partial_{jj}u(x) \pm \mathcal{O}(h_j^3)$ shows

$$\partial_j^{h\pm} u(x) = \partial_j u(x) + \mathcal{O}(h_j^2),$$
$$\partial_j^{h\pm}\partial_k^{h\pm} u(x) = \partial_{jk}u(x) + \mathcal{O}(h_j^2 + h_k^2), \quad j \neq k.$$

An analogous computation for a four-times continuous differentiable function u yields

$$\partial_j^{h+}\partial_j^{h-} u(x) = \partial_{jj}u(x) + \mathcal{O}(h_j^2),$$
$$\partial_j^{h+}\partial_j^{h-}\partial_k^{h+}\partial_k^{h-} u(x) = \partial_{jj}\partial_{kk}u(x) + \mathcal{O}(h_j^2 + h_k^2), \quad j \neq k,$$

and if u is six-times continuously differentiable we have

$$\partial_j^{h+}\partial_j^{h-}\partial_k^{h+}\partial_k^{h-} u(x) = \partial_{jj}\partial_{kk}u(x) + \mathcal{O}(h_j^2 + h_k^2).$$

Thus, for sufficient often continuous differentiable functions we have second order approximations for partial derivatives occuring in the above operators. Replacing the derivatives in (4.17) by appropriate difference operators, we obtain second order approximations. To this end, we define

$$\mathscr{A}_h^{\text{DIFF}} := -\Delta_h := -\sum_{j=1}^d \partial_j^{h+}\partial_j^{h-},$$
(4.19)

and

$$\mathscr{A}_h^{\text{CURV}} := \Delta_h^2 = \sum_{j=1}^d \sum_{k=1}^d \partial_j^{h+}\partial_j^{h-}\partial_k^{h+}\partial_k^{h-}.$$
(4.20)

Analogous, we define $\mathscr{A}_h^{\text{ELAS}}$ by replacing the partial derivative with the approximations above. Then, for example, in two dimensions $\mathscr{A}_h^{\text{ELAS}}$ is given by (cf. (4.18))

$$\mathscr{A}_h^{\text{ELAS}} = -\begin{pmatrix} (2\mu+\lambda)\partial_1^{h+}\partial_1^{h-} + \mu\partial_2^{h+}\partial_2^{h-} & (\lambda+\mu)\partial_1^{h\pm}\partial_2^{h\pm} \\ (\lambda+\mu)\partial_2^{h\pm}\partial_1^{h\pm} & \mu\partial_1^{h+}\partial_1^{h-} + (2\mu+\lambda)\partial_2^{h+}\partial_2^{h-} \end{pmatrix}.$$
(4.21)

Stencil Notation

The application of a difference operator can also be written as a discrete convolution. In particular, for two dimensions it is convenient to define such operators by so-called stencils. Generally, we define a stencil S_h as

$$S_h := [s_k]_h \quad \text{with entries } s_k \in \mathbb{R}, \quad k \in \mathbb{Z}^d.$$
(4.22)

and its application to a function u at $x \in \mathbb{R}^d$

$$S_h u(x) := S_h * u(x) := \sum_{k\in\mathbb{Z}^d} s_k\, u(x - k \odot h).$$
(4.23)

Moreover, for grid points $x_j \in \Omega_h$ (4.23) reads

$$S_h u(x_j) = S_h u_j = \sum_{k\in\mathbb{Z}^d} s_k\, u_{j-k}.$$

For example, in two dimensions the approximation (4.19) of the negative Laplace operator is given by

$$\mathscr{A}_h^{\text{DIFF}} = -\Delta_h = -\begin{bmatrix} & h_2^{-2} & \\ h_1^{-2} & -2(h_1^{-2}+h_2^{-2}) & h_1^{-2} \\ & h_2^{-2} & \end{bmatrix}_h$$

or (4.21) reads

$$
\mathscr{A}_h^{\mathrm{ELAS}} = -\left(
\begin{bmatrix}
\begin{bmatrix}
 & \frac{\mu}{h_2^2} & \\
\frac{2\mu+\lambda}{h_1^2} & -2\left(\frac{2\mu+\lambda}{h_1^2}+\frac{\mu}{h_2^2}\right) & \frac{2\mu+\lambda}{h_1^2} \\
 & \frac{\mu}{h_2^2} &
\end{bmatrix}_h
&
\begin{bmatrix}
-\frac{\mu+\lambda}{4h_1 h_2} & \frac{\mu+\lambda}{4h_1 h_2} \\
\frac{\mu+\lambda}{4h_1 h_2} & -\frac{\mu+\lambda}{4h_1 h_2}
\end{bmatrix}_h \\[2em]
\begin{bmatrix}
-\frac{\mu+\lambda}{4h_1 h_2} & \frac{\mu+\lambda}{4h_1 h_2} \\
\frac{\mu+\lambda}{4h_1 h_2} & -\frac{\mu+\lambda}{4h_1 h_2}
\end{bmatrix}_h
&
\begin{bmatrix}
 & \frac{2\mu+\lambda}{h_2^2} & \\
\frac{\mu}{h_1^2} & -2\left(\frac{\mu}{h_1^2}+\frac{2\mu+\lambda}{h_2^2}\right) & \frac{\mu}{h_1^2} \\
 & \frac{2\mu+\lambda}{h_2^2} &
\end{bmatrix}_h
\end{bmatrix}
\right).
$$

4.4.3 Discretization of Boundary Conditions

After discretizing the domain and differential operators we now turn to the boundary conditions. We proceed similar to the discretization of the derivatives. Therefore, we will derive finite difference approximations for the Neumann boundary conditions

$$\partial_\nu u = 0 \text{ on } \partial(0,1)^d \quad \text{and} \quad \partial_\nu u = \partial_{\nu\nu\nu} u = 0 \text{ on } \partial(0,1)^d$$

and the Dirichlet boundary conditions

$$u = 0 \text{ on } \partial(0,1)^d \quad \text{and} \quad u = \partial_{\nu\nu} u = 0 \text{ on } \partial(0,1)^d.$$

Boundary conditions come into play not only for the modeling of the continuous setting. For the evaluation of a difference operator at the most outer grid points of Ω_h we need function values from points lying outside. Therefore, we extend the known function values from the inner grid points. This is done by reflecting the function at the boundary according to second order finite difference approximations of the imposed boundary conditions.

To make ideas clear we consider the one-dimensional case. This is no loss of generality. Recall from Remarks 4.6 and 4.9, that for $\Omega = (0,1)^d$ the normal derivatives simplify to a single partial derivative. In one dimension, the above boundary conditions read for $u : \mathbb{R} \to \mathbb{R}$

$$u'(0) = u'(1) = 0 \quad \text{and} \quad u'(0) = u'(1) = u'''(0) = u'''(1) = 0$$

and

$$u(0) = u(1) = 0 \quad \text{and} \quad u(0) = u(1) = u''(0) = u''(1) = 0.$$

Furthermore, the discrete analogons of the boundary conditions depend on the particular discretization of the domain. We start with grid-point symmetric discretization.

Figure 4.2: Function values u_k for grid-point symmetric discretization

Grid-Point Symmetry

Let $\Omega = (0,1)$ discretized with $\Omega_h = \{x_k = kh : h = 1/N, \ k = 0, 1, \ldots, N\}$ and $u : \mathbb{R} \to \mathbb{R}$ be a function (cf. Figure 4.2). For the discretization of the derivatives u', u'' and u''' in the boundary conditions we use second order difference schemes. Expanding u in its Taylor series we find

$$u'(x) = \frac{u(x+h) - u(x-h)}{2h} + \mathcal{O}(h^2),$$

$$u''(x) = \frac{u(x+h) - 2u(x) + u(x-h)}{h^2} + \mathcal{O}(h^2),$$

$$u'''(x) = \frac{u(x+2h) - 2u(x+h) + 2u(x-h) - u(x-2h)}{2h^3} + \mathcal{O}(h^2).$$

Thus, the above differences give second order approximations for the derivatives. Neglecting the $\mathcal{O}(h^2)$ term, we have for the Neumann boundary condition at the left boundary $u'(0) = u'(x_0) = 0$, i.e.,

$$\frac{u_1 - u_{-1}}{2h} = 0 \quad \Leftrightarrow \quad u_{-1} = u_1$$

and with the additional requirement $u'''(0) = u'''(x_0) = 0$ we find

$$\frac{u_2 - 2u_1 + 2u_{-1} - u_{-2}}{2h} = 0 \quad \overset{u_{-1}=u_1}{\Leftrightarrow} \quad u_{-2} = u_2.$$

We proceed analogously for the right boundary and Dirichlet conditions. The obtained discrete analogons are summarized in the following table:

Continuous BC	Discrete BC	
	Left ($x = 0$)	Right ($x = 1$)
$u' = 0$	$u_{-1} = u_1$	$u_{N+1} = u_{N-1}$
$u' = u''' = 0$	$u_{-1} = u_1$ and $u_{-2} = u_2$	$u_{N-1} = u_{N+1}$ and $u_{N+2} = u_{N-2}$
$u = 0$	$u_0 = 0$	$u_N = 0$
$u = u'' = 0$	$u_0 = u_0$ and $u_{-1} = -u_1$	$u_N = 0$ and $u_{N+1} = u_{N-1}$

Figure 4.3: Function values u_k for mid-point symmetric discretization

Mid-Point Symmetry

Now, we consider a mid-point symmetric discretization of $\Omega = (0,1)$. Therefore, let $\Omega_h = \{x_k = h/2 + kh : h = 1/N, \ k = 0, 1, \ldots, N\}$ and $u : \mathbb{R} \to \mathbb{R}$ be a function (cf. Figure 4.3). As above, we discretize the derivatives u', u'' and u''' in the boundary conditions with second order difference schemes. Expanding u in its Taylor series we find

$$u'(x) = \frac{u(x + \frac{h}{2}) - u(x - \frac{h}{2})}{h} + \mathcal{O}(h^2)$$

$$u''(x) = \frac{u(x + \frac{3h}{2}) - u(x + \frac{h}{2}) - u(x - \frac{h}{2}) + u(x - \frac{3h}{2})}{2h^2} + \mathcal{O}(h^2).$$

$$u'''(x) = \frac{u(x + \frac{3}{2}) - 3u(x + \frac{h}{2}) + 3u(x - \frac{h}{2}) - u(x - \frac{3h}{2})}{h^3} + \mathcal{O}(h^2).$$

Note that the boundary points 0 and 1 are no grid points. For a discrete formulation of the Dirichlet boundary conditions $u(0) = u(1) = 0$ we use an averaging based on the fact

$$u(x) = \frac{u(x + \frac{h}{2}) + u(x - \frac{h}{2})}{2} + \mathcal{O}(h^2).$$

As above, we the replace derivatives by the difference schemes. Then, for example, the Dirichlet boundary condition $u(0) = 0$ reads in its discrete fashion

$$\frac{u_0 + u_{-1}}{2} = 0 \quad \Leftrightarrow \quad u_{-1} = -u_0$$

and with the additional requirement $u''(0) = 0$

$$\frac{u_{-2} - u_{-1} - u_0 + u_1}{2h^2} = 0 \quad \overset{u_{-1} = -u_0}{\Leftrightarrow} \quad u_{-2} = -u_1.$$

Summarizing, we find:

Continuous BC	Discrete BC	
	Left ($x = 0$)	Right ($x = 1$)
$u' = 0$	$u_{-1} = u_0$	$u_N = u_{N-1}$
$u' = u''' = 0$	$u_{-1} = u_0$ and $u_{-2} = u_1$	$u_{N-1} = u_N$ and $u_{N+2} = u_{N-1}$
$u = 0$	$u_{-1} = -u_0$	$u_N = -u_{-1}$
$u = u'' = 0$	$u_{-1} = -u_0$ and $u_{-2} = -u_1$	$u_N = -u_{N-1}$ and $u_{N+1} = -u_{N-2}$

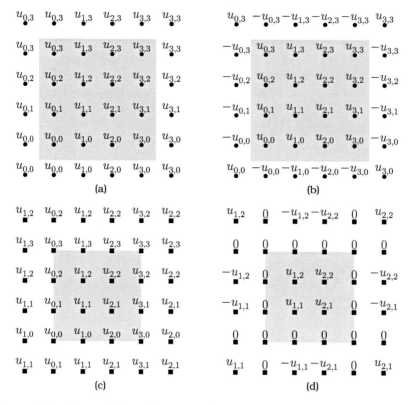

Figure 4.4: Discrete boundary conditions in two dimensions. (a) Neumann BC for mid-point symmetry, (b) Dirichlet BC for mid-point symmetry, (c) Neumann BC for grid-point symmetry, (d) Dirichlet BC for grid-point symmetry

Boundary conditions in higher dimensions

The derived discrete conditions are reflections of function values at the boundary of Ω. More generally, we have patterns for repeating function values as discrete analogons for Dirichlet and Neumann boundary conditions. For grid-point symmetry we have

Neumann: $\dots, u_2, u_1, u_0, u_1, u_2, \dots, u_{N-2}, u_{N-1}, u_N, u_{N-1}, u_{N-2}, \dots$

Dirichlet: $\dots, -u_2, -u_1, 0, u_1, u_2, \dots, u_{N-2}, u_{N-1}, 0, -u_{N-1}, -u_{N-2}, \dots$

and for mesh-point symmetry

Neumann: $\dots, u_1, u_0, u_0, u_1, u_2, \dots, u_{N-2}, u_{N-1}, u_N, u_N, u_{N-1}, \dots$

Dirichlet: $\dots, u_1, u_0, u_0, u_1, u_2, \dots, u_{N-2}, u_{N-1}, u_N, u_N, u_{N-1}, \dots$

In higher dimensions we apply these patterns along the normal directions of the boundary $\partial(0,1)^d$. Figure 4.4 shows the resulting patterns for Dirichlet and Neumann boundary conditions in two dimensions.

4.4.4 Matrix-Vector Notation

With the above discretizations of boundary conditions, we can express the difference equations

$$(\mathrm{id} + \tau \alpha \mathscr{A}_h) u^{n+1}(x_k) = u^n(x_k) - \tau f(x_k, u^n(x_k)) \quad \text{for all } x_k \in \Omega_h$$

in a compact matrix-vector notation. To this end, we collect the values of u and f at the grid points in arrays and vectors, respectively. For easier presentation, we use the same notation for the resulting arrays as for the functions. We define

$$u^n := (u(x_k, t_n))_{x_k \in \Omega_h} \quad \text{and} \quad f^n := (f(x_k, u(x_k, t_n))_{x_k \in \Omega_h}.$$

Furthermore, we derive a matrix A representing the difference operator $\mathscr{A}_h$ such that we obtain a linear system

$$(I + \tau \alpha A) \vec{u}^{n+1} = \vec{u}^n - \tau \vec{f}^n,$$

where $\vec{u}^{n+1}$, $\vec{u}^n$, and $\vec{f}^n$ are associated column vectors to the arrays u^{n+1}, u^n, and f^n.

Before we give a d-dimensional formulation for the matrix representations of the diffusive, curvature and elastic difference operator let us consider the one-dimensional case. In one dimension we have

$$\mathscr{A}_h^{\mathrm{DIFF}} = -\partial^{h+}\partial^{h-} = \frac{1}{h^2}[\,-1\ 2\ -1\,]_h,$$

and $\mathscr{A}_h^{\mathrm{CURV}} = (\mathscr{A}_h^{\mathrm{DIFF}})^2$, $\mathscr{A}_h^{\mathrm{ELAS}} = (\lambda + \mu)\mathscr{A}_h^{\mathrm{DIFF}}$. Since the curvature and elastic operator can be expressed in terms of the diffusive operator, we only consider the matrix representation of $\partial^{h+}\partial^{h-}$. Let $\Omega_h = \{x_0, x_1, \dots, x_N\}$ either a mid-point or grid-point symmetric discretization of the unit interval $(0,1)$ and $u = (u_k)_{k=0}^N \in \mathbb{R}^{N+1}$. Then,

$$\frac{1}{h^2} \begin{pmatrix} a & b & & & \\ 1 & -2 & 1 & & \\ & \ddots & \ddots & \ddots & \\ & & 1 & -2 & 1 \\ & & & c & d \end{pmatrix} \begin{pmatrix} u_0 \\ u_1 \\ \vdots \\ u_{N-1} \\ u_N \end{pmatrix} = \begin{pmatrix} \partial^{h+}\partial^{h-} u_0 \\ \partial^{h+}\partial^{h-} u_1 \\ \vdots \\ \partial^{h+}\partial^{h-} u_{N-1} \\ \partial^{h+}\partial^{h-} u_N \end{pmatrix}$$

where the matrix entries a, b, c, d depend both on the imposed grid symmetry and boundary conditions. For example, if Ω_h is grid point symmetric and Neumann boundary conditions are imposed, then we saw that this reads $u_{-1} = u_1$ and $u_{N+1} = u_{N-1}$. Thus,

$$\partial^{h+}\partial^{h-} u_0 = \frac{u_1 - 2u_0 + u_{-1}}{h^2} = \frac{2u_1 - 2u_0}{h^2}$$

and

$$\partial^{h+}\partial^{h-} u_N = \frac{u_{N+1} - 2u_N + u_{N-1}}{h^2} = \frac{-2u_N + 2u_{N-1}}{h^2}.$$

Hence, in this case we have the entries $a = 2$, $b = -2$, $c = 2$, and $d = -2$.
Defining the matrix

$$D := \begin{pmatrix} -2 & 2 & & & \\ 1 & -2 & 1 & & \\ & \ddots & \ddots & \ddots & \\ & & 1 & -2 & 1 \\ & & & 2 & -2 \end{pmatrix}$$

we obtain the matrix representations

$$A^{\text{DIFF}} = -\tfrac{1}{h^2}D, \quad A^{\text{CURV}} = \tfrac{1}{h^2}D^2, \quad \text{and} \quad A^{\text{ELAS}} = -(2\mu + \lambda)\tfrac{1}{h^2}D,$$

where the square D^2 in the definition of A^{CURV} reads

$$D^2 = \begin{pmatrix} 6 & -8 & 2 & & & & \\ -4 & 7 & -4 & 1 & & & \\ 1 & -4 & 6 & -4 & 1 & & \\ & \ddots & \ddots & \ddots & \ddots & \ddots & \\ & & 1 & -4 & 6 & -4 & 1 \\ & & & 1 & -4 & 7 & -4 \\ & & & & 2 & -8 & 6 \end{pmatrix}.$$

An overview of the matrix representations of $\partial^{h+}\partial^{h-}$ for the different combinations of grid-symmetry and boundary conditions is given in Table 4.2. Moreover, for latter use, the table also lists the matrices for the central difference $\partial^{h\pm}$ needed for the matrix representation of $\mathscr{A}_h^{\text{ELAS}}$ in higher dimensions.

Next, we establish the matrices A^{DIFF}, A^{CURV}, and A^{ELAS} for higher dimensions. We will trace back the d-dimensional case to the one-dimensional case. Therefore, we make use of the Kronecker product and the vec-operation.

The vec-operation reshapes a d-dimensional $N_1 \times N_2 \times \ldots \times N_d$ array to a column vector of length $N = \prod_{\ell=1}^{d} N_\ell$. Therefore, we rearrange the elements of the array by storing them in a row as a column vector. Generally, the vec-operation is given by the relationship

$$\vec{u}_j := \text{vec}(u)_j = u_k \quad :\Leftrightarrow \quad j = k_1 + \sum_{\ell=2}^{d}(k_\ell - 1)\prod_{m=1}^{\ell-1} N_m. \tag{4.24}$$

In particular, in two dimensions arrays are matrices. The vec-operation applied to a matrix $A = (a_{ij}) \in \mathbb{R}^{M \times N}$ takes the elements column-wise and puts them in a row (cf. Example 4.15), that is

$$\vec{A} = (a_{11}, \ldots, a_{M1}; a_{12}, \ldots, a_{M2}, \ldots, a_{1N}, \ldots, a_{MN})^{\top} \in \mathbb{R}^{MN}.$$

Difference		Matrix	BC	Symmetry
$\partial^{h+}\partial^{h-}$	$D_1 :=$	$\begin{pmatrix} -2 & 2 & & & \\ 1 & -2 & 1 & & \\ & \ddots & \ddots & \ddots & \\ & & 1 & -2 & 1 \\ & & & 2 & -2 \end{pmatrix}$	Neumann	grid-point
$\partial^{h+}\partial^{h-}$	$D_2 :=$	$\begin{pmatrix} -2 & 1 & & & \\ 1 & -2 & 1 & & \\ & \ddots & \ddots & \ddots & \\ & & 1 & -2 & 1 \\ & & & 1 & -2 \end{pmatrix}$	Dirichlet	grid-point
$\partial^{h+}\partial^{h-}$	$D_3 :=$	$\begin{pmatrix} -1 & 1 & & & \\ 1 & -2 & 1 & & \\ & \ddots & \ddots & \ddots & \\ & & 1 & -2 & 1 \\ & & & 1 & -1 \end{pmatrix}$	Neumann	mid-point
$\partial^{h+}\partial^{h-}$	$D_4 :=$	$\begin{pmatrix} -3 & 1 & & & \\ 1 & -2 & 1 & & \\ & \ddots & \ddots & \ddots & \\ & & 1 & -2 & 1 \\ & & & 1 & -3 \end{pmatrix}$	Dirichlet	mid-point
$\partial^{h\pm}$	$M_1 :=$	$\begin{pmatrix} 0 & 1 & & & \\ -1 & 0 & 1 & & \\ & \ddots & \ddots & \ddots & \\ & & -1 & 0 & 1 \\ & & & -1 & 0 \end{pmatrix}$	Dirichlet	grid-point
$\partial^{h\pm}$	$M_2 :=$	$\begin{pmatrix} -1 & 1 & & & \\ -1 & 0 & 1 & & \\ & \ddots & \ddots & \ddots & \\ & & -1 & 0 & 1 \\ & & & -1 & -1 \end{pmatrix}$	Dirichlet	mid-point

Table 4.2: Difference matrices

88

The vec-operation is linked to the so-called Kronecker product. The Kronecker product of a $M \times N$ matrix $A = (a_{ij})_{i,j=1}^{M,N}$ and a $P \times Q$ matrix B is the $MP \times NQ$ matrix $A \otimes B$ given by (cf. Example 4.15)

$$A \otimes B := \left(a_{ij} B \right)_{i,j=1}^{M,N}. \tag{4.25}$$

Example 4.15 (vec-Operation and Kronecker Product)
Some examples for the vec-operation applied to matrices.

$$A = \begin{pmatrix} 1 & 3 & 5 \\ 2 & 4 & 6 \end{pmatrix} \quad \Rightarrow \quad \vec{A} = \begin{pmatrix} 1 \\ 2 \\ 3 \\ 4 \\ 5 \\ 6 \end{pmatrix}, \qquad B = \begin{pmatrix} 1 & 4 \\ 2 & 5 \\ 3 & 6 \end{pmatrix} \quad \Rightarrow \quad \vec{B} = \begin{pmatrix} 1 \\ 2 \\ 3 \\ 4 \\ 5 \\ 6 \end{pmatrix},$$

$$C = \begin{pmatrix} 1 & 2 & 3 \\ 4 & 5 & 6 \end{pmatrix} \quad \Rightarrow \quad \vec{C} = \begin{pmatrix} 1 \\ 4 \\ 2 \\ 5 \\ 3 \\ 6 \end{pmatrix}, \qquad D = \begin{pmatrix} 1 & 2 \\ 3 & 4 \\ 5 & 6 \end{pmatrix} \quad \Rightarrow \quad \vec{D} = \begin{pmatrix} 1 \\ 3 \\ 5 \\ 2 \\ 4 \\ 6 \end{pmatrix}.$$

Two examples for the Kronecker product of matrices.

$$\begin{pmatrix} 1 & & \\ & 1 & \\ & & 1 \end{pmatrix} \otimes \begin{pmatrix} 1 & 3 & 5 \\ 2 & 4 & 6 \end{pmatrix} = \left(\begin{array}{ccc|ccc|ccc} 1 & 3 & 5 & & & & & & \\ 2 & 4 & 6 & & & & & & \\ \hline & & & 1 & 3 & 5 & & & \\ & & & 2 & 4 & 6 & & & \\ \hline & & & & & & 1 & 3 & 5 \\ & & & & & & 2 & 4 & 6 \end{array} \right)$$

$$\begin{pmatrix} 1 & 3 & 5 \\ 2 & 4 & 6 \end{pmatrix} \otimes \begin{pmatrix} 1 & & \\ & 1 & \\ & & 1 \end{pmatrix} = \left(\begin{array}{ccc|ccc|ccc} 1 & & & 3 & & & 5 & & \\ & 1 & & & 3 & & & 5 & \\ & & 1 & & & 3 & & & 5 \\ \hline 2 & & & 4 & & & 6 & & \\ & 2 & & & 4 & & & 6 & \\ & & 2 & & & 4 & & & 6 \end{array} \right)$$

We make use of the following link between vec-operation and Kronecker product. Let u, v be $N_1 \times N_2 \times \ldots \times N_d$ arrays, $A^{(\ell)} = (a_{ij}^{(\ell)})_{i,j=1}^{N_\ell, N_\ell}$, $\ell = 1, 2, \ldots, d$ a series of square matrices, and

$$\vec{v} = A^{(d)} \otimes A^{(d-1)} \otimes \ldots \otimes A^{(1)} \cdot \vec{u}.$$

Then, the $j = (j_1, j_2, \ldots, j_d)$-th entry of v is given by

$$v_{j_1 j_2 \ldots j_d} = \sum_{k_d=1}^{N_d} \sum_{k_{d-1}=1}^{N_{d-1}} \cdots \sum_{k_1=1}^{N_1} a_{j_d k_d}^{(d)} \cdot a_{j_{d-1} k_{d-1}}^{(d-1)} \cdots a_{j_1 k_1}^{(1)} \cdot u_{k_1 k_2 \ldots k_d}. \tag{4.26}$$

The proof for this is rather technical and reveals no secrets. So, we will skip it here. For further reading on matrix equations and the Kronecker product see [32, 53].

However, using the above relation we can easily generalize the matrix representations from one dimension to higher dimensions. Let M be one of the difference matrices M_1 or M_2 for $\partial^{h\pm}$ given in Table 4.2. Then the matrix for the partial difference operator $\partial_j^{h\pm}$ in d-dimensions is given by

$$\tfrac{1}{h_j} \left(I_{N_d} \otimes I_{N_{d-1}} \cdots \otimes I_{N_{j+1}} \otimes M \otimes I_{N_{j-1}} \otimes \ldots \otimes I_{N_1} \right),$$

where M is $N_j \times N_j$. Analogous, with D one of the matrices $D_1, \ldots, D_4$ for $\partial^{h+} \partial^{h-}$ from Table 4.2, the partial difference operator $\partial_j^{h+} \partial_j^{h-}$ is represented by

$$\tfrac{1}{h_j} \left(I_{N_d} \otimes I_{N_{d-1}} \cdots \otimes I_{N_{j+1}} \otimes D \otimes I_{N_{j-1}} \otimes \ldots \otimes I_{N_1} \right),$$

where D is $N_j \times N_j$. Replacing the partial difference operators $\partial_j^{h+} \partial_j^{h-}$ and $\partial_j^{h\pm}$ in the definitions of $\mathscr{A}_h^{\mathrm{DIFF}}$, $\mathscr{A}_h^{\mathrm{CURV}}$, and $\mathscr{A}_h^{\mathrm{ELAS}}$ by the corresponding matrices we obtain their d-dimensional matrix representations A^{DIFF}, A^{CURV}, and A^{ELAS}.

To make things clear, next we explicitly give A^{DIFF}, A^{CURV}, and A^{ELAS} for the important cases of two and three dimensions. In the following we omit the indication of dimensions for the matrices taking part in Kronecker products. If nothing else is said, the matrix at the ℓ-th position is assumed to be $N_\ell \times N_\ell$ (counted from right to left).

In two dimensions the matrix for $\mathscr{A}_h^{\mathrm{DIFF}}$ reads

$$A^{\mathrm{DIFF}} = -I_2 \otimes \left(\tfrac{1}{h_1^2}(I \otimes D) + \tfrac{1}{h_2^2}(D \otimes I) \right)$$

$$= - \begin{pmatrix} \tfrac{1}{h_1^2}(I \otimes D) + \tfrac{1}{h_2^2}(D \otimes I) & \\ & \tfrac{1}{h_1^2}(I \otimes D) + \tfrac{1}{h_2^2}(D \otimes I) \end{pmatrix},$$

the matrix for $\mathscr{A}_h^{\mathrm{CURV}}$ is given by

$$A^{\mathrm{CURV}} = (A^{\mathrm{DIFF}})^2 = I_2 \otimes \left(\tfrac{1}{h_1^2}(I \otimes D) + \tfrac{1}{h_2^2}(D \otimes I) \right)^2$$

$$= \begin{pmatrix} \left(\tfrac{1}{h_1^2}(I \otimes D) + \tfrac{1}{h_2^2}(D \otimes I) \right)^2 & \\ & \left(\tfrac{1}{h_1^2}(I \otimes D) + \tfrac{1}{h_2^2}(D \otimes I) \right)^2 \end{pmatrix},$$

and $\mathscr{A}_h^{\mathrm{ELAS}}$ is represented by

$$A^{\mathrm{ELAS}} = - \begin{pmatrix} \tfrac{2\mu+\lambda}{h_1^2}(I \otimes D) + \tfrac{\mu}{h_2^2}(D \otimes I) & \tfrac{\mu+\lambda}{h_1 \cdot h_2}(M \otimes M) \\ \tfrac{\mu+\lambda}{h_1 \cdot h_2}(M \otimes M) & \tfrac{\mu}{h_1^2}(I \otimes D) + \tfrac{2\mu+\lambda}{h_2^2}(D \otimes I) \end{pmatrix}.$$

In three dimensions we obtain

$$A^{\text{DIFF}} = -I_3 \otimes \left(\tfrac{1}{h_1^2}(I \otimes I \otimes D) + \tfrac{1}{h_2^2}(I \otimes D \otimes I) + \tfrac{1}{h_3^2}(D \otimes I \otimes I) \right),$$

$$A^{\text{CURV}} = I_3 \otimes \left(\tfrac{1}{h_1^2}(I \otimes I \otimes D) + \tfrac{1}{h_2^2}(I \otimes D \otimes I) + \tfrac{1}{h_3^2}(D \otimes I \otimes I) \right)^2,$$

$$A^{\text{ELAS}} = - \begin{pmatrix} A_{11}^{\text{ELAS}} & A_{12}^{\text{ELAS}} & A_{13}^{\text{ELAS}} \\ A_{12}^{\text{ELAS}} & A_{22}^{\text{ELAS}} & A_{23}^{\text{ELAS}} \\ A_{13}^{\text{ELAS}} & A_{23}^{\text{ELAS}} & A_{33}^{\text{ELAS}} \end{pmatrix}$$

with

$$A_{11}^{\text{ELAS}} = \tfrac{2\mu+\lambda}{h_1^2}(I \otimes I \otimes D) + \tfrac{\mu}{h_2^2}(I \otimes D \otimes I) + \tfrac{\mu}{h_3^2}(D \otimes I \otimes I),$$

$$A_{22}^{\text{ELAS}} = \tfrac{\mu}{h_1^2}(I \otimes I \otimes D) + \tfrac{2\mu+\lambda}{h_2^2}(I \otimes D \otimes I) + \tfrac{\mu}{h_3^2}(D \otimes I \otimes I),$$

$$A_{33}^{\text{ELAS}} = \tfrac{\mu}{h_1^2}(I \otimes I \otimes D) + \tfrac{\mu}{h_2^2}(I \otimes D \otimes I) + \tfrac{2\mu+\lambda}{h_3^2}(D \otimes I \otimes I),$$

$$A_{12}^{\text{ELAS}} = \tfrac{\mu+\lambda}{h_1 h_2}(I \otimes M \otimes M),$$

$$A_{13}^{\text{ELAS}} = \tfrac{\mu+\lambda}{h_1 h_3}(M \otimes I \otimes M),$$

$$A_{23}^{\text{ELAS}} = \tfrac{\mu+\lambda}{h_2 h_3}(M \otimes M \otimes I).$$

Note that matrices D and M for A^{ELAS} must be properly chosen for Dirichlet boundary conditions with the same grid symmetry.

Summarizing, we have $N_1 \times N_2 \ldots \times N_d \times d$ arrays u^n, u^{n+1} and f^n collecting the function values at the grid points. Then, reshaping them to column vectors $\vec{u}^{n+1}$, $\vec{u}^n$, and $\vec{f}^n$ of length $(\prod_{\ell=1}^d N_\ell) \cdot d$, the discretized PDE for the gradient flow reads

$$(I + \tau \alpha A)u^{n+1} = u_n - \tau f^n,$$

with A one of the derived matrices for the diffusive, curvature, or elastic smoother.

4.5 Numerics

For the computation of a numerical solution to the gradient flow (4.15) we proceed in general as follows.

- Choose an initial displacement $u^0 \in \mathbb{R}^{N_1 \times \ldots \times N_d \times d}$, $\tau > 0$ and $\alpha > 0$.

- for $n = 0, 1, 2, \ldots$ do

    1. Compute forces $f^n = f(x_k, u(x_k, n\tau))$

    2. Compute new displacement u^{n+1} by solving the linear system

    $$(I + \tau \alpha A)u^{n+1} = u^n - \tau f^n.$$

- end

We have to solve two major numerical problems. The first one is the computation of the forces and the second one is solving the linear system. For the solution to the first problem, in section 4.5.1 we present a fast method for the computation of the forces. Using recently developed algorithms for non-equispaced Fourier transforms, we end up with $\mathcal{O}(N)$ algorithm to evaluate the forces at $N = N_1 N_2 \cdots N_d$ points.

To compute a step of the gradient flow we have to solve a linear system $I + \tau \alpha A$ with A the matrix for the diffusive, curvature, or elastic smoother. The systems are rather large, so that fast solvers are inevitable. The dimension of $I + \tau \alpha A$ is $Nd \times Nd$ with $N = \#\Omega_h$ the number of pixels/voxels. Therefore, the dimensions grow rapidly. Typical image sizes in medical applications are, e.g., 512×512 and $256 \times 256 \times 256$ yielding hundreds of thousands and millions of unknowns. The following table illustrates the rapid growth of the number of unknowns with the image size.

image size	128^2	256^2	512^2	128^3	256^3	512^3
#unknowns	32.768	131.072	524.288	6.291.456	50.331.648	402.653.184

The above linear systems are sparse, structured with constant coefficients, and the equations stem from so-called elliptic PDEs. In particular, for the diffusive and curvature registration there are fast methods for direct solving. Using real Fourier transforms, we show how to compute an exact solution with $\mathcal{O}(N \log N)$ operations.

Another state-of-the-art method for solving systems originating from an elliptic PDE is geometric multigrid. Multigrid is an iterative method requiring $\mathcal{O}(N)$ operations which applies to all three kinds of registration. However, we start with the computation of the forces. In section 4.5.3 we show how to solve the linear systems for the diffusive and curvature registrations directly and section 4.5.4 is devoted to multigrid methods.

4.5.1 Computation of Forces

In this section we discuss the computation of the array f^n containing the forces from the array u^n containing the displacements on the underlying grid Ω_h. We will present a fast algorithm with linear arithmetic complexity. The method is based on using truncated Fourier series as kernel functions for approximating densities. This allows for the application of recently developed non-equispaced fast Fourier transforms [35, 36], yielding an $\mathcal{O}(N)$ algorithm with $N = \#\Omega_h$.

For unification, in the following let Ω_h contain all grid points x_k with grid spacing $h = (1/N_1, 1/N_2, \ldots, 1/N_d)$ and $k \in \mathcal{I}$, where $\mathcal{I}$ is an index set. That is $\Omega_h := \{x_k : k \in \mathcal{I}\}$ with the index set

$$\mathcal{I} := \{k = (k_1, k_2, \ldots, k_d) \in \mathbb{Z}^d : k_\ell = 1, 2, \ldots, N_\ell, \ \ell = 1, 2, \ldots, d\}.$$

Then defining $x_k := -h/2 + k \odot h$ yields midpoint symmetry and defining $x_k := k \odot h$ yields grid-point symmetry. By this, both

$$f^n = (f^n_k)_{k \in \mathcal{I}} \quad \text{and} \quad u^n = (u^n_k)_{k \in \mathcal{I}}$$

are $N_1 \times N_2 \times \ldots \times N_d \times d$ arrays with components $f^n_k \in \mathbb{R}^d$ the force and $u^n_k \in \mathbb{R}^d$ the displacement at the k-th grid-point x_k, respectively. Moreover, for images $R, T \in \text{Img}(\Omega)$ we define

$$R_k := R(x_k) \quad \text{and} \quad T^n_k := T(x_k - u^n_k).$$

With this notation, the forces for the direct approach using approximating densities (cf. equation (4.6)) reads

$$f^n_k = L^\sigma_{RT^n}(R_k, T^n_k) \, \nabla T(x_k - u^n_k)$$

and for the approximating approach (cf. Lemma 4.3)

$$f^n_k = (K_\sigma * L^\sigma_{RT^n})(R_k, T^n_k) \, \nabla T(x_k - u^n_k)$$

where

$$L^\sigma_{RT^n}(r, t) := \frac{1}{|\Omega|} \left(\frac{\partial_2 p^\sigma_{RT^n}(r, t)}{p^\sigma_{RT^n}(r, t)} - \frac{p^{\sigma\prime}_{T^n}(t)}{p^\sigma_{T^n}(t)} \right).$$

The crucial point in the computation of the forces is the weighting $L^\sigma_{RT^n}$ and $K_\sigma * L^\sigma_{RT^n}$, respectively. In practice, images, in the sense of continuous functions in $\text{Img}(\Omega)$, are obtained by interpolating some discrete data. Thus, the values $T(x_k - u^n_k)$ and $\nabla T(x_k - u^n_k)$ are computed by an interpolation scheme. Standard approaches use, e.g., linear interpolation to compute $T(x_k - u^n_k)$ and approximate the gradient using finite differences [30, 18, 17, 19], or images are represented as cubic B-Splines and the derivatives are explicitly known [58, 56, 57]. We will not discuss this any further and turn to the computation of the leading weighting term.

Here, we prefer the direct approach since we have no need for a convolution of $L^\sigma_{RT^n}$ with the kernel K_σ. Neglecting the convolution makes no essential difference. The convolution introduces an additional smoothness to the $L^\sigma_{RT^n}$, controlling the length of the gradient ∇T. This effect is similar to the one obtained by choosing σ large and therefore resulting very smooth approximating densities.

In [30] Hermosillo proposes to proceed as follows for the computation of the forces in the approximating approach:

1. Introduce bins $B_k =]k\beta - \frac{\beta}{2}, k\beta + \frac{\beta}{2}]$ with bin-size $\beta > 0$ and bin-centers $(r_i, t_j) := (i\beta, j\beta)$ of $B_i \times B_j$ (cf. Examples 2.7 and 2.8). Compute the joint histogram

$$H_{ij} = \#\{x_k \in \Omega_h \; : \; R(x_k) \in B_i \text{ and } T(x_k - u^n_k) \in B_j\}/\#\Omega_h,$$

   i.e., count the number of intensities from the images on Ω_h falling into each bin divided by the total number of gridpoints.

```
function L = computeForceWeighting(R,T,sigma,nbins)
h = 0.5/(nbins-1);
N = prod(size(R));
Rscaled = 1 + round( (R+1/4)* (2*nbins-1) );
Tscaled = 1 + round( (T+1/4)* (2*nbins-1) );
H = zeros(nbins,nbins);
for k=1:N
    H(Rscaled(k),Tscaled(k)) = H(Rscaled(k),Tscaled(k)) + 1/N;
end;
t   = linspace(-1/4,1/4,2*nbins);
K   = 1/sqrt(2*pi*sigma^2)*exp(-t.^2 /(2*sigma^2) );
p1  = conv2(sum(H,1),K,'same');
dp1 = 1/(2*h)*(p1([2:end end]) - p1([1 1:end-1]));
p2  = conv2(conv2(H,K,'same'),K','same');
dp2 = 1/(2*h)*(p2(:,[2:end end]) - p2(:,[1 1:end-1]));
dE  = dp2./p2 - repmat(dp1./p1,nbins,1);
KL  = h^2*conv2(conv2(dE,K,'same'),K','same');
L   = interp2(KL',Rscaled,Tscaled,'linear')';
```

Figure 4.5: Matlab code example for computing $K_\sigma * L_{RT^n}^\sigma$ as proposed in [30]

2. Approximate $p_{RT^n}^\sigma(r_i, t_j)$ and $p_{T^n}^\sigma(t_j)$ by a discrete convolution of the histogram with K_σ (in particular in [30] with the Gaussian).

3. Compute the derivatives $\partial_2 p_{RT^n}^\sigma(r_i, t_j)$ and $p_{T^n}^{\sigma\prime}(t_j)$ using central finite differences.

4. Compute $L_{RT^n}^\sigma(r_i, t_j)$ and subsequently $(K_\sigma * L_{RT^n}^\sigma)(r_i, t_j)$ by a discrete convolution with K_σ (in particular with K_σ a Gaussian).

5. Evaluate $(K_\sigma * L_{RT^n}^\sigma)(R_k, T_k^n)$ by linear interpolation.

This method is easy to implement and has a computational complexity of $\mathcal{O}(N)$ with N the number of gridpoints and pixels/voxels, respectively. Figure 4.5.1 shows a straightforward implementation in Matlab.

Drawbacks are that the whole procedure is discontinuous by the nature of histograms and therefore still not differentiable. Moreover, the approximation is based on artificially introduced points (r_i, t_j) and the needed values at (R_k, T_k^n) are once more an approximation obtained by interpolation. Figure 4.6 illustrates how the binning causes steps in the weighting term $K_\sigma * L_{RT^n}^\sigma$ of the forces.

In [26] we showed this proceeding might cause misalignments and proposed a different method without using any bins. Instead of computing $L_{RT^n}^\sigma$ at artificial points we compute the required values directly. As proposed in [26] we make use of recently developed fast algorithms for

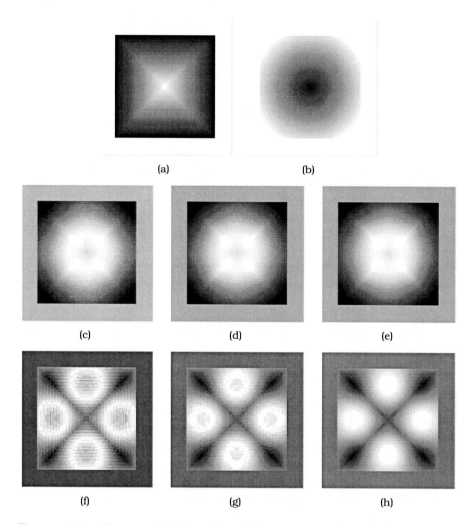

Figure 4.6: Force weighting $K_\sigma * L^\sigma_{RT^n}$ computed with histograms:
(a) reference, (b) template, (c)-(e) force weighting for $\sigma = 1/20$
and #bins=32,64,256, (f)-(h) force weighting for $\sigma = 1/50$ and
#bins=32,64,256

computing *discrete Gauss transforms* defined as

$$G(z) := \sum_{j=1}^{N} \alpha_j \exp(-c\|z - z_j\|^2) \quad \text{with} \quad c > 0, \quad \alpha_j \in \mathbb{C}, \quad z_j \in \mathbb{R}^d. \quad (4.27)$$

For the computation of $L^\sigma_{RT^n}(R_k, T^n_k)$ we have to evaluate $p^\sigma_{T^n}(T^n_k)$, $p^{\sigma'}_{T^n}(T^n_k)$ and $p^\sigma_{RT^n}(R_k, T^n_k)$, $\partial_2 p^\sigma_{RT^n}(R_k, T^n_k)$. To this end we approximate the integral over the domain Ω using the common midpoint rule by summing over the points in Ω_h. For $p^\sigma_{T^n}(T^n_k)$ we obtain

$$p^\sigma_{T^n}(T^n_k) = \frac{1}{|\Omega|} \int_\Omega K_\sigma(T^n_k - T(x - u^n(x)))\, dx$$

$$\approx \frac{\prod_{j=1}^{d} h_j}{|\Omega|} \sum_{\Omega_h} K_\sigma(T^n_k - T(x - u^n(x_k)))$$

$$= \frac{\prod_{j=1}^{d} h_j}{|\Omega|} \sum_{\ell \in \mathcal{I}} K_\sigma(T^n_k - T^n_\ell)$$

and analogously for $p^{\sigma'}_{T^n}(T^n_k)$, $p^\sigma_{RT^n}(R_k, T^n_k)$, $\partial_2 p^\sigma_{RT^n}(R_k, T^n_k)$ we have

$$p^{\sigma'}_{T_{u^n}}(T^n_k) \approx \frac{\prod_{j=1}^{d} h_j}{|\Omega|} \sum_{\ell \in \mathcal{I}} K'_\sigma(T^n_k - T^n_\ell),$$

$$p^\sigma_{RT_{u^n}}(R_k, T^n_k) \approx \frac{\prod_{j=1}^{d} h_j}{|\Omega|} \sum_{\ell \in \mathcal{I}} K_\sigma(R_k - R_\ell)K_\sigma(T^n_k - T^n_\ell),$$

$$p^{\sigma'}_{T_{u^n}}(T^n_k) \approx \frac{\prod_{j=1}^{d} h_j}{|\Omega|} \sum_{\ell \in \mathcal{I}} K_\sigma(R_k - R_\ell)K'_\sigma(T^n_k - T^n_\ell).$$

Thus, we obtain

$$L^\sigma_{RT^n}(R_k, T^n_k) \approx \frac{\sum_{\ell \in \mathcal{I}} K_\sigma(R_k - R_\ell)K'_\sigma(T^n_k - T^n_\ell)}{\sum_{\ell \in \mathcal{I}} K_\sigma(R_k - R_\ell)K_\sigma(T^n_k - T^n_\ell)} - \frac{\sum_{\ell \in \mathcal{I}} K'_\sigma(T^n_k - T^n_\ell)}{\sum_{\ell \in \mathcal{I}} K_\sigma(T^n_k - T^n_\ell)}. \quad (4.28)$$

Clearly, choosing K_σ as a Gaussian the above approximations reveal as discrete Gauss transforms at $z = T^n_k$ and $z = (R_k, T^n_k)$, respectively with $c = -\frac{1}{2\sigma^2}$ and $\alpha_j = \frac{\prod_{\ell=1}^{d} h_\ell}{|\Omega|\sigma}$, $z_j = T^n_j$ and $\alpha_j = \frac{\prod_{\ell=1}^{d} h_\ell}{|\Omega|\sigma^2}$, $z_j = (R_j, T^n_j)$, respectively. If we want to evaluate the Gauss transform at a single point, we have to compute $N = \#\mathcal{I}$ summands yielding a computational complexity of $\mathcal{O}(N)$ operations. Additionally, we have to evaluate the transform at T^n_k and (R_k, T^n_k) for all $k \in \mathcal{I}$, and therefore at N points. Neither the values for R_k nor the values T^n_k are regular in any sense, i.e., they are values from a regular grid. Therefore, we cannot take any advantages for the computation of the differences $R_k - R_\ell$ and $T^n_k - T^n_\ell$ and it is inevitable to compute them for all combinations $(k, \ell) \in \mathcal{I} \times \mathcal{I}$. Thus, an exact implementation for the Gauss transform requires $\mathcal{O}(N^2)$ operations. The number of gridpoints N is quite large so that a direct computation is much too time consuming. For example, computing the Gauss transforms for 2D images with $256 \times 256 = 65536$ pixels with an off-the-shelf PC computer performs within minutes (cf. [37]).

To speed up the computation we employ a so-called fast Gauss transform that approximates the discrete Gauss transform with arbitrary precision. Several methods for the fast Gauss transform have been developed [21, 37]. Here we follow [26, 37] using truncated Fourier series for the kernel K_σ. Applying non-equispaced fast Fourier transforms (NFFT) we end up with an $\mathcal{O}(N)$ algorithm for the computation of the forces again.

The method described below is quite general and not restricted to a particular kernel K_σ [46]. Nevertheless, in the following we give explicit expressions for the Gaussian, yielding the fast Gauss transform proposed in [26, 37].

We start with the definition of truncated Fourier series of kernels. For K_σ we define its truncated Fourier series

$$K_{\sigma,n}(x) := \sum_{\ell=-\frac{n}{2}}^{\frac{n}{2}-1} a_\ell \, e^{\mathrm{i}2\pi\ell x}, \quad a_\ell = \left\langle K_\sigma, e^{-\mathrm{i}2\pi\ell\cdot} \right\rangle_{L^2(\mathbb{R})}, \quad n \in \mathbb{N} \text{ even}.$$

In particular for the Gaussian $K_\sigma(x) = \frac{1}{\sqrt{2\pi}\sigma} e^{-\frac{x^2}{2\sigma^2}}$ the coefficients a_ℓ are given by

$$a_\ell = \left\langle K_\sigma, e^{-\mathrm{i}2\pi\ell\cdot} \right\rangle_{L^2(\mathbb{R})} = \int_{-\infty}^{\infty} K_\sigma(x) \, e^{-\mathrm{i}2\pi\ell x} \, \mathrm{d}x = e^{-2\pi^2\ell^2\sigma^2},$$

Note that $K_{\sigma,n}$ is a periodic function with $K_{\sigma,n}(x) = K_{\sigma,n}(x+k)$ for all $x \in \mathbb{R}$ and $k \in \mathbb{Z}$. To this end, we restrict ourselves to a single period $-\frac{1}{2} \leq x \leq \frac{1}{2}$. This is no loss of generality, since the arguments of K_σ are intensity values and differences of intensity values, respectively. By definition, the range of an image is bounded, such that an appropriate scaling of the images to the range $[-\frac{1}{4}, \frac{1}{4}]$ ensures $-\frac{1}{2} \leq R(x) - R(y) \leq \frac{1}{2}$ and $-\frac{1}{2} \leq T(x) - T(y) \leq \frac{1}{2}$ for all $x, y \in \mathbb{R}^d$.

However, if $K_\sigma \in L^2(\mathbb{R})$ and therefore the coefficients a_ℓ are well defined, then $K_{\sigma,n}$ converges to a periodic version of K_σ, that is

$$K_{\sigma,n}(x) \quad \to \quad \sum_{k\in\mathbb{Z}} K_\sigma(x-k) \quad \text{as } n \to \infty, \quad x \in \mathbb{R}.$$

Thus, for well localized K_σ, e.g., by choosing σ small, or even compactly supported in the interval $[-\frac{1}{2}, \frac{1}{2}]$, $K_{\sigma,n}$ provides a close approximation to K_σ. Note that n must be chosen depending on σ in order to avoid Gibbs phenomena (cf. Figure 4.5.1). Loosely speaking, the smaller σ, the larger n must be chosen. For an exact error estimate in the particular case of K_σ the Gaussian see [37].

Furthermore, we are interested in the computation of the derivative of the kernel. In the case of $K_{\sigma,n}$ the derivative is easily obtained by a slight

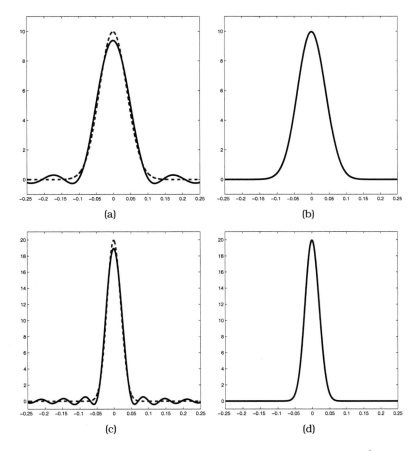

Figure 4.7: $K_{n,\sigma}(x)$ (solid) and the Gaussian $K_\sigma(x) = \frac{1}{\sqrt{2\pi\sigma^2}}e^{-\frac{x^2}{2\sigma^2}}$ (dashed) with $\sigma = 0.04$ for (a) $n = 16$, (b) $n = 32$, and $\sigma = 0.02$ for (c) $n = 32$, (d) $n = 64$.

modification of the coefficients a_ℓ. Differentiating $K_{\sigma,n}$ w.r.t. x yields

$$K'_{\sigma,n}(x) = \sum_{\ell=-\frac{n}{2}}^{\frac{n}{2}-1} \underbrace{a_\ell(\mathrm{i}2\pi\ell)}_{=:a'_\ell} e^{\mathrm{i}2\pi\ell x} = \sum_{\ell=-\frac{n}{2}}^{\frac{n}{2}-1} a'_\ell e^{\mathrm{i}2\pi\ell x}.$$

Next, we plug $K_{\sigma,n}$ and $K'_{\sigma,n}$ into the above sums for the computation of the density approximations (4.28). Then we can compute them with *non-equispaced discrete Fourier transforms*. .

Given nodes $v_k \in \mathbb{R}^d$, $k = 1, 2, \ldots, N$ and $n \in \mathbb{N}$ even, the non-equispaced discrete Fourier transform (NDFT) of a vector is given by the multiplication with the matrix

$$F = \left(e^{-\mathrm{i}2\pi\ell^\top v_k}\right)_{\substack{k=1,2,\ldots,N, \\ \ell\in\{-\frac{n}{2},\ldots,\frac{n}{2}-1\}^d}} \in \mathbb{C}^{n^d \times N}.$$

The arithmetic complexity of the NDFT is $\mathcal{O}(n^d N)$. Analogous to discrete Fourier transform (DFT) and fast Fourier transform (FFT), there is a fast algorithm for computing the NDFT, too - the non-equispaced fast Fourier transform (NFFT). Instead of $\mathcal{O}(n^d N)$ operation, the NFFT reduces the computational cost to $\mathcal{O}(N + n^d \log n)$ [35, 36].

Moreover, there are three additional DFTs and NFFTs, respectively. These are multiplications with the transposed $F^\top$, the conjugated $\overline{F}$, and the hermitian $F^H = \overline{F}^\top$ of F. Here, we only need the transposed and conjugated transform. For the sake of completeness we define all four NDFT transforms. They are given by

$$\text{NDFT} : \mathbb{C}^{n^d} \to \mathbb{C}^N, \quad x \mapsto Fx,$$
$$\text{NDFT}^\top : \mathbb{C}^N \to \mathbb{C}^{n^d}, \quad x \mapsto F^\top x,$$
$$\overline{\text{NDFT}} : \mathbb{C}^{n^d} \to \mathbb{C}^N, \quad x \mapsto \overline{F}x,$$
$$\text{NDFT}^H : \mathbb{C}^N \to \mathbb{C}^{n^d}, \quad x \mapsto F^H x = \overline{F}^\top x.$$

where the conjugated $\overline{F}$ of F is given by

$$\overline{F} = \left(e^{i2\pi \ell^\top v_k} \right)_{\substack{k=1,2,\dots,N, \\ \ell \in \{-\frac{n}{2},\dots,\frac{n}{2}-1\}^d}} \in \mathbb{C}^{n^d \times N}.$$

Note that $\overline{F}$ is the matrix F for the nodes $-v_k$, $k = 1, 2, \dots, N$. For a free implementation of those NDFTs and their fast versions as NFFTs in the C programming language see [35].

Each of the four sums in (4.28) can be computed for all R_k and T_k^n by a NDFT$^\top$ and a subsequent $\overline{\text{NDFT}}$. For $\sum_{j \in \mathcal{I}} K_{\sigma,n}(T_k^n - T_j^n)$ we have

$$\sum_{j \in \mathcal{I}} K_{\sigma,n}(T_k^n - T_j^n) = \sum_{j \in \mathcal{I}} \sum_{\ell=-\frac{n}{2}}^{\frac{n}{2}-1} a_\ell \, e^{i2\pi\ell(T_k^n - T_j^n)}$$

$$= \sum_{\ell=-\frac{n}{2}}^{\frac{n}{2}-1} a_\ell \underbrace{\left(\sum_{j \in \mathcal{I}} e^{-i2\pi\ell T_j^n} \right)}_{\text{NDFT}^\top} e^{i2\pi\ell T_k^n} = \underbrace{\sum_{\ell=-\frac{n}{2}}^{\frac{n}{2}-1} a_\ell b_\ell \, e^{i2\pi\ell T_k^n}}_{\overline{\text{NDFT}}}$$

with $b_\ell := \sum_{j \in \mathcal{I}} e^{-i2\pi\ell T_j^n}$. Analogously we obtain

$$\sum_{j \in \mathcal{I}} K'_{\sigma,n}(T_k^n - T_j^n) = \sum_{\ell=-\frac{n}{2}}^{\frac{n}{2}-1} a'_\ell b_\ell \, e^{i2\pi\ell T_k^n},$$

$$\sum_{j \in \mathcal{I}} K_{\sigma,n}(R_k - R_j) K_{\sigma,n}(T_k^n - T_j^n) = \sum_{\ell_1=-\frac{n}{2}}^{\frac{n}{2}-1} \sum_{\ell_2=-\frac{n}{2}}^{\frac{n}{2}-1} a_{\ell_1} a_{\ell_2} b_{\ell_1 \ell_2} \, e^{i2\pi(\ell_1 R_k + \ell_2 T_k^n)},$$

$$\sum_{j \in \mathcal{I}} K_{\sigma,n}(R_k - R_j) K'_{\sigma,n}(T_k^n - T_j^n) = \sum_{\ell_1=-\frac{n}{2}}^{\frac{n}{2}-1} \sum_{\ell_2=-\frac{n}{2}}^{\frac{n}{2}-1} a_{\ell_1} a'_{\ell_2} b_{\ell_1 \ell_2} \, e^{i2\pi(\ell_1 R_k + \ell_2 T_k^n)}$$

where $b_{\ell_1 \ell_2} = \sum_{j \in \mathcal{I}} e^{-i2\pi(\ell_1 R_k + \ell_2 T_k^n)}$, as the coefficients b_ℓ, can be computed via NDFT$^\top$.

```
function y = ndft1d(v,n,type,x)
k = [-n/2:n/2-1]';
F = exp(-i*2*pi*v*k');
switch type,
    case 'normal'    , y = F*x;
    case 'conjugated', y = conj(F)*x;
    case 'transposed', y = conj(F)'*x;
    case 'adjoint'   , y = F'*x;
end;

function y = ndft2d(v1,v2,n,type,x)
[K1,K2] = ndgrid(-n/2:n/2-1);
k1       = reshape(K1,n^2,1);
k2       = reshape(K2,n^2,1);
F        = exp(-i*2*pi*(v1*k1'+v2*k2'));
switch type,
    case 'normal'    , y = F*x;
    case 'conjugated', y = conj(F)*x;
    case 'transposed', y = conj(F)'*x;
    case 'adjoint'   , y = F'*x;
end;

function L = computeForceWeighting(R,T,sigma,n)
N = prod(size(R));
v1 = reshape(R,N,1);
v2 = reshape(T,N,1);
e = ones(N,1);
k = [-n/2:n/2-1]';
a = exp(-2*pi^2*sigma^2*k.^2); a(1)=0; da = (i*2*pi*k).*a;
A = reshape(a*a',n^2,1); dA = reshape(a*conj(da)',n^2,1);
b = ndft1d(v2,n,'transposed',e);
B = ndft2d(v1,v2,n,'transposed',e);
p1   = real( ndft1d(v2,n,'conjugated',a.*b) );
dp1  = real( ndft1d(v2,n,'conjugated',da.*b) );
p2   = real( ndft2d(v1,v2,n,'conjugated',A.*B) );
dp2  = real( ndft2d(v1,v2,n,'conjugated',dA.*B) );
vecL = dp2./p2 - dp1./p1;
L = reshape(vecL,size(R));
```

Figure 4.8: Matlab code example for computing $L^\sigma_{RT^n}$ using NDFTs

Summarizing, we obtain the following algorithm:
Let F_1 be the NDFT matrix to the nodes $\{T^n_k\}_{k\in\mathcal{I}}$, F_2 be the NDFT matrix to the nodes $\{(R_k, T^n_k)\}_{k\in\mathcal{I}}$, and $o_M := (1, 1, \ldots, 1)^\top \in \mathbb{R}^M$.

1. Compute the coefficient vector $a = (a_1, a_2, \ldots, a_n)^\top$ for $K_{\sigma,n}$, e.g., set $a_\ell = e^{-2\pi^2\sigma^2\ell^2}$ yielding the fast Gauss transform.

2. Compute the coefficients $a' = (a'_1, a'_2, \ldots, a'_n)^\top$, with $a'_\ell = (i2\pi\ell)a_\ell$ for the derivative $K'_{\sigma,n}$.

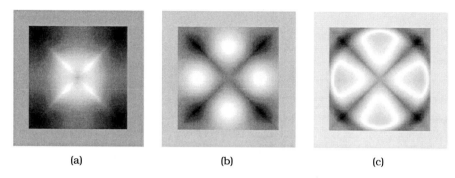

Figure 4.9: Force weighting $L_{RT^n}^\sigma$ computed with NFFT's: (a)-(c) force weighting for $\sigma = 1/10, 1/20, 1/40$

3. Compute $b_1 = F^\top o_n$ and $b_2 = F_2^\top o_{n^2}$.

4. Compute $\mathrm{vec}(p_1) = \overline{F}_1 \left(a \odot b_1 \right)$ and $\mathrm{vec}(p_1') = \overline{F}_1 \left(a' \odot b_1 \right)$

5. Compute $\mathrm{vec}(p_2) = \overline{F}_2 \left(\mathrm{vec}(aa^\top) \odot b_2 \right)$ and $\mathrm{vec}(p_2') = \overline{F}_2 \left(\mathrm{vec}(aa'^\top) \odot b_2 \right)$

6. Compute $L_k = \frac{(p_2')_k}{(p_2)_k} - \frac{(p_1')_k}{(p_1)_k}$

Then L is an approximation of $L_{RT^n}^\sigma$ given in (4.28), i.e., $L_{RT^n}^\sigma(R_k, T_k^n) \approx L_k$. Finally, using the NFFT for computing the multiplications with the NDFT matrices F_1, F_2 the arithmetic complexity for this method is $\mathcal{O}(N + n^d \log n)$. Furthermore, n only depends on fixed chosen σ. Thus, n is a constant and the overall asymptotic complexity comes to $\mathcal{O}(N)$. As already mentioned, to avoid Gibbs phenomena, the value of n grows with smaller σ. Gained from empirical experiences, n stays moderate. In all experiments done, values for n in the range between 32 and 128 were sufficient.

An example for a possible implementation in Matlab is given in Figure 4.5.1. In order to present a working code, the NDFT is given explicitly as matrix vector multiplication. Clearly, the functions ndft1d and ndft2d should be replaced by fast NFFT routines, e.g., using the C library available from [35]. Nevertheless, even when using the NDFT routines, the asymptotic complexity for fixed σ and n, respectively, is $\mathcal{O}(N)$, too.

Figure 4.9 displays the force weighting for the image pair shown in Figure 4.6 computed with the NFFT method. In contrast to the histogram based method we obtain a smooth function without discontinuities. Furthermore, in the approximative approach $L_{RT^n}^\sigma$ is additionally smoothed by a convolution with K_σ. As mentioned in the beginning a similar smoothing effect is obtained in the direct approach by choosing σ large. In the particular case for K_σ the Gaussian we have the well-known relation for the convolution of two kernels $K_\sigma * K_\tau = K_{\sigma+\tau}$. Therefore, $K_\sigma * L_{RT^n}^\sigma$ is roughly comparable to $L_{RT^n}^{2\sigma}$, cf. Figure 4.6(e) displaying $K_\sigma * L_{RT^n}^\sigma$ for $\sigma = 1/20$ and 4.9(a) displaying $L_{RT^n}^\sigma$ for $\sigma = 1/10$.

Summarizing, we obtain an elegant method for computing a continuous differentiable force weighting at the same level of asymptotic complexity as the histogram based method.

4.5.2   Choice of σ

So far we have not discussed the choice of the parameter σ for the approximating densities. Figure 4.6 and Figure 4.9 illustrate how different values for σ change the force weighting. Roughly speaking, the smaller σ is, the more "attention" is payed to small structures in the images. Nevertheless, this observation does not tell us how to choose σ.

A common way to choose σ automatically is using so-called *cross-validation techniques* [24, 30, 51, 62]. These techniques are particularly used for kernel density estimation (cf. section 3.2) to find a reasonable approximation to the density. Cross-validation is crucial when only few data is available. We will not discuss those techniques here. For more see, e.g., [51]. Nevertheless, we mention two (for us) important points. First, cross-validation is an optimization method w.r.t. σ with a non-neglectable computational cost. Second, it is based on the assumption of independent identically distributed data. Here, we derived an approximation for the joint density of images that does not meet the assumption of independent identically distributed data. Therefore, we cannot apply cross-validation directly. Furthermore, we are using quite large data, such that reasonable values for σ could be chosen roughly.

Another way to choose σ comes from its interpretation as a scaling parameter. From Lemma 3.2 we have the important relations

$$p_{RT}^{\sigma} = K_{\sigma} * p_{RT} \qquad \text{and} \qquad p_{RT}^{\sigma} \to p_{RT} \text{ as } \sigma \to 0$$

if p_{RT} exists and is continuous. Therefore, choosing σ large neglects small structures which have rare intensities, such that the image registration in general emphasizes the alignment of large structures in the images.

Summarizing, there is no general or preferable method for choosing σ. Cross-validation techniques allow for an automatic choice to obtain a close approximation to the "true" density. Treating σ as a scaling parameter might speed up the registration, but usually relies on heuristics.

4.5.3   Direct Solver for Linear Systems

In the following we will show how to compute an exact solution of the linear system

$$(I + \tau \alpha A)\vec{u}^{n+1} = \vec{u}^n - \tau \vec{f}^n \quad \Leftrightarrow \quad \vec{u}^{n+1} = (I + \tau \alpha A)^{-1}(\vec{u}^n - \tau \vec{f}^n)$$

for the diffusive registration with $A = A^{\text{DIFF}}$ and curvature registration with $A = A^{\text{CURV}}$. The heart of the method are discrete trigonometric transforms. They can be computed efficiently such that we obtain a fast solvers with a computational complexity of $\mathcal{O}(N \log N)$.

To make ideas clear and to avoid notational overhead we consider the two-dimensional case. The presented principles can be easily extended to arbitrary dimensions. To this end, we finally also give explicit expressions for three dimensions as well.

First of all, we notice that in contrast to elastic registration, the linear systems $I + \tau\alpha A^{\text{DIFF}}$ and $I + \tau\alpha A^{\text{CURV}}$ decouple. For $N_1 \times N_2 \ldots N_d \times d$ arrays u^{n+1}, u^n and f^n, the vectorized versions read

$$\vec{u}^{n+1} = (\vec{u}_1^{n+1}, \vec{u}_2^{n+1}, \ldots, \vec{u}_d^{n+1})^\top,$$
$$\vec{u}^{n} = (\vec{u}_1^{n}, \vec{u}_2^{n}, \ldots, \vec{u}_d^{n})^\top,$$
$$\vec{f}^{n} = (\vec{f}_1^{n}, \vec{f}_2^{n}, \ldots, \vec{f}_d^{n})^\top$$

where u_ℓ^{n+1}, u_ℓ^n, f_ℓ^n are $N_1 \times N_2 \ldots \times N_d$ arrays holding the displacements and the forces for the ℓ-th dimension. Therefore, the linear system $(I + \tau\alpha A)\vec{u}^{n+1} = \vec{u}^n - \tau f^n$ reads

$$\begin{pmatrix} I + \tau\alpha A_{1,1} & & & \\ & I + \tau\alpha A_{2,2} & & \\ & & \ddots & \\ & & & I + \tau\alpha A_{d,d} \end{pmatrix} \begin{pmatrix} \vec{u}_1^{n+1} \\ \vec{u}_2^{n+1} \\ \vdots \\ \vec{u}_d^{n+1} \end{pmatrix} = \begin{pmatrix} \vec{u}_1^n - \tau\vec{f}_1^n \\ \vec{u}_2^n - \tau\vec{f}_2^n \\ \vdots \\ \vec{u}_d^n - \tau\vec{f}_d^n \end{pmatrix}$$

$$\Leftrightarrow \quad (I + \tau\alpha A_0)u_\ell^{n+1} = u_\ell^n - \tau f_\ell^n \quad \text{for } \ell = 1, 2, \ldots, d$$

where $A_0 := A_{1,1} = A_{2,2} = \ldots = A_{d,d}$. Thus, instead of solving the large $Nd \times Nd$ system we can sequentially solve the smaller $N \times N$ systems separately. For example for the diffusive registration in two dimensions with $A = A^{\text{DIFF}}$ we have $A_{1,1}^{\text{DIFF}} = A_{2,2}^{\text{DIFF}} = \frac{1}{h_1^2}(I \otimes D) + \frac{1}{h_2^2}(D \otimes I) =: A_0^{\text{DIFF}}$ and

$$(I + \tau\alpha A_0^{\text{DIFF}})u_1^{n+1} = u_1^n - \tau f_1^n, \quad \text{and} \quad (I + \tau\alpha A_0^{\text{DIFF}})u_2^{n+1} = u_2^n - \tau f_2^n.$$

Next, we show how to solve these $N \times N$ systems directly. As already mentioned, the key of the method are of trigonometric transforms - discrete sine (DST) and cosine transforms (DCT). Both, the DCT and DST are special kinds of Fourier transforms involving only real arithmetic [54]. There is more than one sine and cosine transform. Following the nomenclature of [20, 55] in the following we will use the standard transforms of types I,II, and III. However, the DCTs and DSTs are real linear mappings

$$\text{DCT} : \mathbb{R}^N \to \mathbb{R}^N, \quad x \mapsto Cx \quad \text{and} \quad \text{DST} : \mathbb{R}^N \to \mathbb{R}^N, \quad x \mapsto Sx.$$

The matrices C and S for one-dimensional transforms are listed in Table 4.3. The matrices for transformations in higher dimensions are given by

DCT-I $\quad C^I = \frac{1}{\sqrt{2(N-1)}} \left(\varepsilon_k \cos \frac{jk\pi}{N-1} \right)_{j,k=0}^{N-1,N-1}$, $\quad \varepsilon_k = \begin{cases} 1, & k = 0, N-1, \\ 2, & k = 1, \dots, N-2. \end{cases}$

DCT-II $\quad C^{II} = \frac{1}{\sqrt{2N}} \left(2 \cos \frac{(j+\frac{1}{2})k\pi}{N} \right)_{j,k=0}^{N-1,N-1}$

DCT-III $\quad C^{III} = \frac{1}{\sqrt{2N}} \left(\varepsilon_k \cos \frac{j(k+\frac{1}{2})\pi}{N} \right)_{j,k=0}^{N-1,N-1}$, $\quad \varepsilon_k = \begin{cases} 1, & k = 0, \\ 2, & k = 1, \dots, N-1. \end{cases}$

DST-I $\quad S^I = \frac{1}{\sqrt{2(N+1)}} \left(2 \sin \frac{jk\pi}{N+1} \right)_{j,k=1}^{N,N}$

DST-II $\quad S^{II} = \frac{1}{\sqrt{2N}} \left(2 \sin \frac{j(k-\frac{1}{2})\pi}{N} \right)_{j,k=1}^{N,N}$

DST-III $\quad S^{III} = \frac{1}{\sqrt{2N}} \left(\varepsilon_k \sin \frac{(j-\frac{1}{2})k\pi}{N} \right)_{j,k=1}^{N,N}$, $\quad \varepsilon_k = \begin{cases} 2, & k = 1, \dots, N-1, \\ 1, & k = N. \end{cases}$

Table 4.3: Standard sine and cosine transformations I-III

Kronecker products. For example the standard DCT-II of a $N_1 \times N_2 \times \dots \times N_d$ array x reads

$$\mathrm{DCT}(\vec{x}) = C\vec{x} \quad \text{where } C = C^{II} \otimes C^{II} \otimes \dots \otimes C^{II}.$$

All transforms can be computed with $\mathcal{O}(N \log N)$ operations [8, 20, 61]. An implementation in the C-programming language is available at [20]. Our method for solving the linear systems is based on eigenvalue decompositions of the difference matrices given in Table 4.2. All of them can be diagonalized with the sine and cosine matrices given in Table 4.3.

Theorem 4.16 (Decomposition of Difference Matrices)
Let D_1, D_2, D_3, D_4 be the difference matrices from Table 4.2, S^I, S^{II}, S^{III} the sine and, C^I, C^{II}, C^{III} the cosine matrices from Table 4.3. Furthermore let $\lambda_j^N := -2 + 2\cos\frac{j\pi}{N}$. Then
a) $D_1 = C^I \operatorname{diag} \left(\lambda_{j-1}^{N-1} \right)_{j=1}^{N} C^I$, $\quad C^I C^I = I$,
b) $D_2 = S^I \operatorname{diag} \left(\lambda_j^{N+1} \right)_{j=1}^{N} S^I$, $\quad S^I S^I = I$,
c) $D_3 = C^{III} \operatorname{diag} \left(\lambda_{j-1}^{N-1} \right)_{j=1}^{N} C^{II}$, $\quad C^{III} C^{II} = I$,
d) $D_4 = S^{III} \operatorname{diag} \left(\lambda_j^N \right)_{j=1}^{N} S^{II}$, $\quad S^{III} S^{II} = I$.

Proof. The above identities are well-known and they are straightforward to verify. Nevertheless, proving the statements is rather technical and we will skip it here. For a proof see [55, 54]. ∎

Note, that all eigenvalues are non-negative, so that all difference matrices are at least positive semi-definite. In particular D_1 and D_3 have a zero eigenvalue and are therefore positive semi-definite, whereas D_2 and D_4 are positive definite.

We start with outlining the principle for solving in two dimensions. In general A^{DIFF} and A^{CURV} are sums from Kronecker products of identity matrices I with a difference matrix D. The above theorem shows D can be decomposed into $V^{-1}\Lambda V$, where Λ is a diagonal matrix. Thus, we have for the diffusive registration

$$
\begin{aligned}
A_0^{\mathrm{DIFF}} &= -(\tfrac{1}{h_1^2}I \otimes D + \tfrac{1}{h_2^2}D \otimes I) \\
&= -(\tfrac{1}{h_1^2}(V^{-1}V) \otimes (V^{-1}\Lambda V) + \tfrac{1}{h_2^2}(V^{-1}\Lambda V) \otimes (V^{-1}V)) \\
&= -(V^{-1} \otimes V^{-1})\,(\tfrac{1}{h_1^2}I \otimes \Lambda + \tfrac{1}{h_2^2}\Lambda \otimes I)\,(V \otimes V) \\
&= W^{-1}\Sigma W
\end{aligned}
$$

with $W := V \otimes V$ and $\Sigma := -(\tfrac{1}{h_1^2}I \otimes \Lambda + \tfrac{1}{h_2^2}\Lambda \otimes I)$. Furthermore, for the matrix $I \otimes \Lambda \in \mathbb{R}^{N_1 N_2 \times N_1 N_2}$ with $\Lambda = \mathrm{diag}(\lambda_1,\ldots,\lambda_{N_1})$ and $I \in \mathbb{R}^{N_2 \times N_2}$, from (4.26) we have the representation $I \otimes \Lambda = \mathrm{diag}\,\mathrm{vec}(\lambda_{j_1})_{j_1,j_2=1}^{N_1,N_2}$. Analogous, with $\Lambda = \mathrm{diag}(\lambda_1,\ldots,\lambda_{N_2})$ equation (4.26) implies $\Lambda \otimes I = \mathrm{diag}\,\mathrm{vec}(\lambda_{j_2})_{j_1,j_2=1}^{N_1,N_2}$. Therefore, Σ is a diagonal matrix given by

$$
\Sigma = \mathrm{diag}\,\mathrm{vec}\,(\sigma_{j_1 j_2})_{j_1,j_2=1}^{N_1,N_2} \quad \text{where} \quad \sigma_{j_1 j_2} = -\left(\frac{\lambda_1}{h_1^2} + \frac{\lambda_2}{h_2^2}\right). \tag{4.29}
$$

This representation is quite useful, since we will compute multiplications of Σ with vectorized $N_1 \times N_2$ arrays $u = (u_{j_1 j_2})_{j_1,j_2=1}^{N_1,N_2}$. Since Σ is a diagonal matrix it scales $\vec{u}$ component-wise. With the above representation of Σ the scaling reveals as

$$
\Sigma \vec{u} = \mathrm{vec}\,(\sigma_{j_1 j_2} u_{j_1 j_2})_{j_1,j_2=1}^{N_1,N_2}.
$$

Furthermore, $W = V \otimes V$ and $W^{-1} = V^{-1} \otimes V^{-1}$ are matrices for two-dimensional sine and cosine transforms, respectively. As already mentioned, a multiplication can be efficiently computed $\mathcal{O}(N \log N)$ where $N = N_1 N_2$.

However, with W and Σ we obtain

$$
I + \tau \alpha A_0^{\mathrm{DIFF}} = W^{-1}(I + \tau \alpha \Sigma)W = W^{-1}\Gamma^{\mathrm{DIFF}}W
$$

where

$$
\Gamma^{\mathrm{DIFF}} := (I + \tau \alpha \Sigma) = \mathrm{diag}\,\mathrm{vec}\,(1 + \tau \alpha \sigma_{j_1 j_2})_{j_1,j_2=1}^{N_1,N_2}
$$

Then the inverse of $I + \tau \alpha A_0^{\mathrm{DIFF}}$ is given by

$$
(I + \tau \alpha A_0^{\mathrm{DIFF}})^{-1} = W^{-1}(I + \tau \alpha \Sigma)^{-1}W = W^{-1}(\Gamma^{\mathrm{DIFF}})^{-1}W.
$$

Analogous, for the curvature registration with $A_0^{\mathrm{CURV}} = (A_0^{\mathrm{DIFF}})^2$ we obtain the representation

$$
A_0^{\mathrm{CURV}} = (A_0^{\mathrm{DIFF}})^2 = (W^{-1}\Sigma W)(W^{-1}\Sigma W) = W^{-1}\Sigma^2 W
$$

and therefore

$$I + \tau \alpha A_0^{\mathrm{CURV}} = W^{-1}(I + \tau \alpha \Sigma^2)W = W^{-1}\Gamma^{\mathrm{CURV}}W$$

with

$$\Gamma^{\mathrm{CURV}} := (I + \tau \alpha \Sigma^2) = \mathrm{diag\,vec}\left(1 + \tau \alpha \sigma_{j_1 j_2}^2\right)_{j_1, j_2 = 1}^{N_1, N_2}.$$

Thus, the inverse of $I + \tau \alpha A_0^{\mathrm{CURV}}$ is given by

$$(I + \tau \alpha A_0^{\mathrm{CURV}})^{-1} = W^{-1}(I + \tau \alpha \Sigma^2)^{-1}W = W^{-1}(\Gamma^{\mathrm{CURV}})^{-1}W.$$

Summarizing, in two dimensions we solve the linear system

$$(I + \tau \alpha A)\vec{u}^{n+1} = \vec{u}^n - \vec{f}^n$$

with $A = A^{\mathrm{DIFF}}$ or $A = A^{\mathrm{CURV}}$ for $u^{n+1} = (u_1^{n+1}, u_2^{n+1})$ and the right-hand-side $u^n - \tau f^n = (u_1^n - \tau f_1^n, u_2^n - \tau f_2^n)$ by

$$\vec{u}_1^{n+1} = (W^{-1}\Gamma^{-1}W)\,(\vec{u}_1^m - \tau \vec{f}_1^n) \quad \text{and} \quad \vec{u}_2^{n+1} = (W^{-1}\Gamma^{-1}W)\,(\vec{u}_2^m - \tau \vec{f}_2^n),$$

with either $\Gamma = \Gamma^{\mathrm{DIFF}}$ or $\Gamma = \Gamma^{\mathrm{CURV}}$. Since Γ is a diagonal matrix its inverse Γ^{-1} is given by simply inverting the diagonal entries of Γ and therefore diagonal, too. A multiplication of a vector with $\Gamma^{-1} \in \mathbb{R}^{N \times N}$ costs $\mathcal{O}(N)$ arithmetic operations. Both, the multiplication with W and W^{-1} are two-dimensional sine and cosine transforms that can be computed with $\mathcal{O}(N \log N)$ operations [20], so that the overall computational complexity is $\mathcal{O}(N \log N)$.

Figure 4.10 shows a simple implementation for solving in two dimensions for the case of Neumann boundary conditions and mid-point symmetric discretization (cf. Theorem 4.16 and Table 4.4). The multiplications with $W = V \otimes V$ and $W^{-1} = V^{-1} \otimes V^{-1}$ are computed by the methods dct3 and dct2. To this end, we use the well-known identity for the Kronecker product $B \otimes A \,\mathrm{vec}(X) = \mathrm{vec}(AXB^\top)$ [53, 32]. Clearly, the presented implementation involves matrix multiplication and therefore has a computational complexity of $\mathcal{O}(N^2)$. To this end, the methods dct3 and dct2 should be replaced by fast versions, e.g., by using the implementation in the C-programming language from [20].

Now we have seen how to compute a solution in the two-dimensional case. The above analysis can be carried out directly for arbitrary dimensions. For example, in three dimensions, the matrices W and W^{-1} are given by

$$W = V \otimes V \otimes V \quad \text{and} \quad W^{-1} = V^{-1} \otimes V^{-1} \otimes V^{-1},$$

```
function v=dct2(u);
[n1,n2] = size(u);
j = [0:n1-1]';   k = j;
Cn1 = 1/sqrt(2*n1) * 2 * cos(j*(k'+1/2)*pi/n1);
j = [0:n2-1]';   k = j;
Cn2 = 1/sqrt(2*n2) * 2 * cos(j*(k'+1/2)*pi/n2);
v = Cn1 * u * Cn2';

function v=dct3(u);
[n1,n2] = size(u);
j = [0:n1-1]';   k = j;
Cn1 = 1/sqrt(2*n1) * cos((j+1/2)*k'*pi/n1) * diag([1 2*ones(1,n1-1)]);
j = [0:n2-1]';   k = j;
Cn2 = 1/sqrt(2*n2) * cos((j+1/2)*k'*pi/n2) * diag([1 2*ones(1,n2-1)]);
v = Cn1 * u * Cn2';

function [u1new,u2new] = solveCurvature(h,tau,alpha,u1,u2,f1,f2)
[n1,n2] = size(u1);
[J1,J2] = ndgrid(0:n1-1,0:n2-1);
Lambda1  = -2 + 2*cos(J1*pi/(n1-1));
Lambda2  = -2 + 2*cos(J2*pi/(n2-1));
Sigma    = -( 1/h(1)^2*Lambda1 + 1/h(2)^2*Lambda2 );
invGamma = 1./(1+tau*alpha*Sigma.^2);
u1new = dct3(invGamma.*dct2(u1-tau*f1));
u2new = dct3(invGamma.*dct3(u2-tau*f2));
```

Figure 4.10: Matlab code example for solving $(I + \tau\alpha A^{\mathrm{CURV}})\vec{u}^{n+1} = \vec{u}^n - \tau\vec{f}^n$ in two dimensions with Neumann boundary conditions and mid-point symmetric discretization.

and Σ reads

$$\Sigma = \operatorname{diag} \operatorname{vec}\left(\sigma_{j_1 j_2 j_3}\right)_{j_1 j_2 j_3=1}^{N_1, N_2, N_3} \quad \text{with} \quad \sigma_{j_1 j_2 j_3} = -\left(\frac{\lambda_{j_1}}{h_1^2} + \frac{\lambda_{j_2}}{h_2^2} + \frac{\lambda_{j_3}}{h_3^2}\right).$$

Then

$$I + \tau\alpha A_0^{\mathrm{DIFF}} = W^{-1}(I + \tau\alpha\Sigma)W \quad \text{and} \quad I + \tau\alpha A_0^{\mathrm{CURV}} = W^{-1}(I + \tau\alpha\Sigma^2)W,$$

and, analogous to the above, we solve the linear system in three dimensions for $u^{n+1} = (u_1^{n+1}, u_2^{n+1}, u_3^{n+1})$ and the right-hand-side $u^n - \tau f^n = (u_1^n - \tau f_1^n, u_2^n - \tau f_2^n, u_3^n - \tau f_3^n)$ by

$$\vec{u}_1^{n+1} = (W^{-1}\Gamma^{-1}W)(\vec{u}_1^n - \tau\vec{f}_1^n),$$
$$\vec{u}_2^{n+1} = (W^{-1}\Gamma^{-1}W)(\vec{u}_2^n - \tau\vec{f}_2^n),$$
$$\vec{u}_3^{n+1} = (W^{-1}\Gamma^{-1}W)(\vec{u}_3^n - \tau\vec{f}_3^n)$$

with

$$\Gamma = \Gamma^{\mathrm{DIFF}} = I + \tau\alpha\Sigma = \operatorname{diag} \operatorname{vec}\left(1 + \tau\alpha\sigma_{j_1 j_2 j_3}\right)_{j,1 j_2 j_3=1}^{N_1, N_2, N_3}$$

for diffusive registration or

$$\Gamma = \Gamma^{\mathrm{CURV}} = I + \tau\alpha\Sigma^2 = \mathrm{diag\,vec}\left(1 + \tau\alpha\sigma^2_{j_1 j_2 j_3}\right)^{N_1,N_2,N_3}_{j_1,j_2,j_3=1}$$

for curvature registration.

The following theorem summarizes the resulting expressions for arbitrary dimensions. The proof is a direct extension of the above analysis presented for two dimensions. Nevertheless, we will skip it here.

Theorem 4.17 (Inversion of $I + \tau\alpha A_0$)

Let D be one of the difference matrices given in Table 4.2, with Eigenvalue decomposition $D = V_\ell^{-1}\Lambda V_\ell \in \mathbb{R}^{N_\ell \times N_\ell}$, where $\Lambda = \mathrm{diag}(\lambda_j)^{N_\ell}_{j=1}$ and $\lambda_j \geq 0$. Furthermore, let $W := V_d \otimes \ldots \otimes V_2 \otimes V_1$, and

$$\Sigma := \mathrm{diag\,vec}(\sigma_{j_1 j_2 \cdots j_d})^{N_1,N_2,\ldots,N_d}_{j_1 j_2 \cdots j_d=1} \quad where \quad \sigma_{j_1 j_2 \cdots j_d} = -\sum_{\ell=1}^{d}\frac{\lambda_{j_\ell}}{h_\ell^2}.$$

Then $\Gamma^{\mathrm{DIFF}} := I + \tau\alpha\Sigma$ and $\Gamma^{\mathrm{CURV}} := I + \tau\alpha\Sigma^2$ are invertible for all $\tau,\alpha > 0$, such that

$$(I + \tau\alpha A_0^{\mathrm{DIFF}})^{-1} = W^{-1}(\Gamma^{\mathrm{DIFF}})^{-1}W, \quad (I + \tau\alpha A_0^{\mathrm{CURV}})^{-1} = W^{-1}(\Gamma^{\mathrm{CURV}})^{-1}W.$$

Based on the theorem, the general procedure for solving $(I + \tau\alpha A)\vec{u}^{n+1} = \vec{u}^n - \tau\vec{f}^n$ in d-dimensions where $u^{n+1} = (u_1^{n+1},\ldots,u_d^{n+1})$, $u^n = (u_1^n,\ldots,u_d^n)$, and $f^n = (f_1^n,\ldots,f_d^n)$ reads as follows:

- for $\ell = 1,2,\ldots,d$ do

  1. Compute $\vec{v} = W(\vec{u}_\ell^n - \tau\vec{f}_\ell^n)$ by a d-dimensional DCT and DST, respectively.   (forward transform)

  2. Multiply with the eigenvalues of $(I + \tau\alpha A_0)^{-1}$, that is

  $$w_{j_1 j_2 \cdots j_d} = \begin{cases} (1 + \tau\alpha\sigma_{j_1 j_2 \cdots j_d})^{-1}\,v_{j_1 j_2 \cdots j_d} & \text{for } A_0 = A_0^{\mathrm{DIFF}}, \\ (1 + \tau\alpha\sigma^2_{j_1 j_2 \cdots j_d})^{-1}\,v_{j_1 j_2 \cdots j_d} & \text{for } A_0 = A_0^{\mathrm{CURV}}. \end{cases}$$

  (inversion)

  3. Compute $\vec{u}_\ell^{n+1} = W^{-1}\vec{w}$ by a d-dimensional DCT and DST, respectively.   (backward transform)

- end for

Note that $\sigma_{j_1 j_2 \cdots j_d}$, the matrix W for the forward transform, and the matrix W^{-1} for the backward transform depend on the imposed boundary conditions and grid symmetry. An overview of the valid transforms and eigenvalues $\sigma_{j_1 j_2 \cdots j_d}$ for the combinations of the several boundary conditions and grid symmetries is given in Table 4.4.

Boundary Conditions	Symmetry	backward transform	$\sigma_{j_1 j_2 \dots j_d}$	forward transform
Neumann	grid-point	DCT-I	$\displaystyle\sum_{\ell=1}^{d} \frac{-2 + 2\cos\frac{(j_\ell-1)\pi}{N-1}}{h_\ell^2}$	DCT-I
Dirichlet	grid-point	DST-I	$\displaystyle\sum_{\ell=1}^{d} \frac{-2 + 2\cos\frac{j_\ell\pi}{N+1}}{h_\ell^2}$	DST-I
Neumann	mid-point	DCT-III	$\displaystyle\sum_{\ell=1}^{d} \frac{-2 + 2\cos\frac{(j_\ell-1)\pi}{N-1}}{h_\ell^2}$	DCT-II
Dirichlet	mid-point	DST-III	$\displaystyle\sum_{\ell=1}^{d} \frac{-2 + 2\cos\frac{j_\ell\pi}{N}}{h_\ell^2}$	DST-II

$(I + \tau\alpha A_0)^{-1} = W^{-1}\Gamma^{-1}W$ with backward transform W^{-1}, forward transform W,

$$\Gamma^{-1} = \operatorname{diag\,vec}\left((1 + \tau\alpha\sigma_{j_1 j_2 \dots j_d})^{-1}\right)_{j_1,j_2 \dots j_d=1}^{N_1,N_2,\dots,N_d} \quad \text{for diffusive registration, and}$$

$$\Gamma^{-1} = \operatorname{diag\,vec}\left((1 + \tau\alpha\sigma_{j_1 j_2 \dots j_d}^2)^{-1}\right)_{j_1,j_2 \dots j_d=1}^{N_1,N_2,\dots,N_d} \quad \text{for curvature registration.}$$

Table 4.4: Eigenvalue decomposition of $(I + \tau\alpha A_0)^{-1}$

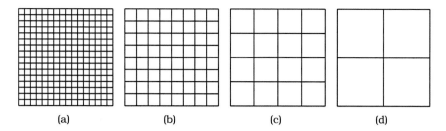

(a)   (b)   (c)   (d)

Figure 4.11: Grid hierarchy

4.5.4 Geometric Multigrid for Linear Systems

In this section we show how to solve the linear system

$$(I + \tau\alpha A)\vec{u}^{n+1} = \vec{u}^n - \tau\vec{f}^n$$

by a multigrid method. To this end we consider the formulation

$$(\mathrm{id} + \tau\alpha\mathscr{A}_h)u^{n+1}(x_k) = u^n(x_k) - \tau f(x_k, u^n(x_k)) \quad \text{for all } x_k \in \Omega_h \qquad (4.30)$$

where the difference operator $\mathscr{A}_h$ is in stencil notation.

Multigrid is a general method to solve linear systems and there are several components and options one might choose. However, a detailed analysis and exploiting suitable components and options is beyond the scope of this thesis. In the following, we briefly outline the basic components of a multigrid method. Subsequently, we explicitly consider standard components for solving our linear systems in two dimensions. For further reading refer to the books [3, 25, 59].

In the following, we deal with a domain $\Omega \subset \mathbb{R}^d$ and grids $\Omega_h, \Omega_{2h}, \Omega_{4h}, \ldots$ with different coarsening (cf. Figure 4.11). To this end, we introduce some notation.

Let $\Omega \subset \mathbb{R}^d$ be a domain, and $\Omega_h = \{x_k \ : \ k \in \mathbb{Z}^d\} \subset \mathbb{R}^d$ be a grid. Then we call a function restricted to Ω_h a *grid function* and define the set of all grid functions by

$$\mathcal{G}(\Omega_h; \mathbb{R}^k) := \{u^h = (u(x_k))_{x_k \in \Omega_h} \ : \ u : \mathbb{R}^d \to \mathbb{R}^k\}.$$

To keep notation short, in the following we also write $\mathcal{G}(\Omega_h)$ for $\mathcal{G}(\Omega_h; \mathbb{R}^k)$ when the range $\mathbb{R}^k$ either emanates from the context or does not play a crucial role.

Clearly, $\mathcal{G}(\Omega_h; \mathbb{R}^k)$ can be identified as the $\mathbb{R}^{N \times k}$ with $N = \#\Omega_h$ the number of grid points. As above, to keep notation short, we write $u_k^h = u(x_k)$ for $x_k \in \Omega_h$. Defining the difference operator $\mathscr{L}_h := \mathrm{id} + \tau\alpha\mathscr{A}_h$, and the grid functions

$$u^h = (u^{n+1}(x_k))_{x_k \in \Omega_h} \quad \text{and} \quad g^h = (u^n(x_k) - \tau f(x_k, u^n(x_k)))_{x_k \in \Omega_h},$$

the equation (4.30) reads

$$\mathcal{L}_h u^h = g^h.$$

Furthermore, let u^h be the solution of $\mathcal{L}_h u^h = g^h$ and v^h an approximation to u^h. Then the *error* e^h is defined as

$$e^h := u^h - v^h,$$

and the *residual* by

$$r^h := g^h - \mathcal{L}^h v^h.$$

For every linear operator $\mathcal{L}_h$, we have the important relation of error and residual

$$\mathcal{L}_h e^h = \mathcal{L}_h(u^h - v^h) = \mathcal{L}_h u^h - \mathcal{L}_h v^h = g^h - \mathcal{L}_h v^h = r^h.$$

The equation $\mathcal{L}_h e^h = r^h$ is the so-called *residual equation.*
Next, we briefly introduce the basic components for multigrid methods. First of all, we have to transfer function values between different grids with different grid spacings. This is done by interpolation.
The interpolation from a fine grid to a coarser grid is called *restriction*. In particular, here we define the restriction operator

$$I_h^{2h} : \mathcal{G}(\Omega_h) \to \mathcal{G}(\Omega_{2h})$$

interpolating the fine grid with spacing h to the coarser grid with doubled grid spacing $2h$. Vice versa, the interpolation from a coarse to a finer grid is called *prolongation*. Analogous to the restriction operator I_h^{2h}, we define the prolongation operator

$$I_{2h}^h : \mathcal{G}(\Omega_{2h}) \to \mathcal{G}(\Omega_h)$$

Finally, we need a so-called smoothing operator depending on $\mathcal{L}_h$

$$\mathcal{S} : \mathcal{G}(\Omega_h) \times \mathcal{G}(\Omega_h) \to \mathcal{G}(\Omega_h), \quad (u^h, g^h) \mapsto \mathcal{S}(u^h, g^h).$$

Furthermore, we define its ν-times repeated application to grid functions $u, g \in \mathcal{G}(\Omega_h)$ as

$$\mathcal{S}^\nu(u, g) := \mathcal{S}(u^{\nu-1}, g) \quad \text{with} \quad u^{\nu-1} = \mathcal{S}(u^{\nu-2}, g) \text{ and } u^0 := u.$$

Common smoothers are relaxation schemes such as Jacobi or Gauss-Seidel relaxation [59]. Later on, we use a smoother based on Jacobi relaxation for solving our linear systems.
Based on these components, we proceed as follows for solving $\mathcal{L}_h u^h = g^h$ for a given right-hand-side g^h and initial guess v^h. First, we apply the smoother ν_1 times, $v^h \leftarrow \mathcal{S}^{\nu_1}(v^h, g^h)$ and subsequently compute the

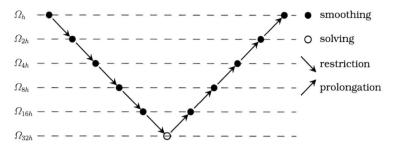

Ω_h • smoothing

Ω_{2h} ○ solving

Ω_{4h} ↘ restriction

Ω_{8h} ↗ prolongation

Ω_{16h}

Ω_{32h}

Figure 4.12: V-Cycle with coarsest grid spacing $h_{\max} = 32h$

residual $r^h = g^h - \mathscr{L}_h v^h$. Next, we solve the residual equation on the next coarser grid with doubled grid spacing $2h$. To this end, we restrict the residual to coarser grid and compute $r^{2h} = I_h^{2h} r^h$ and compute e^{2h} by solving the residual equation $\mathscr{L}_{2h} e^{2h} = r^{2h}$ on the coarser grid. After solving we interpolate the error back to the fine grid. That is, we compute e^h by prolongation $e^h = I_{2h}^h e^{2h}$. Then we correct the approximate solution with the error, $v^h \leftarrow v^h + e^h$ and finally apply the smoother ν_2 times again. For solving the coarse grid residual equation $\mathscr{L}_{2h} e^{2h} = r^{2h}$ we proceed in the same way, so that we obtain a recursive procedure. The recursion terminates when we reach a given coarsest grid. On the coarsest grid we solve the equation $\mathscr{L}_h v^h = g^h$ directly. This can be done, for example, numerically by exhaustive smoothing.

The whole procedure is called multigrid V-cycle. Its name stems from the diagram shown in Figure 4.12 that looks like the character "V". Summarizing, the described procedure reads as follows.

$v^h = $ V-CYCLE(ν_1, ν_2, v^h, g^h)

- if $h = h_{\max}$ then $v^h \leftarrow \mathscr{L}_h^{-1} g^h$ (solve)

- else

     1. $v^h \leftarrow \mathscr{S}^{\nu_1}(v^h, g^h)$ (pre-smooth)

     2. $r^h \leftarrow g^h - \mathscr{L}_h v^h$ (compute fine grid residual)

     3. $r^{2h} \leftarrow I_h^{2h} r^h$ (compute coarse grid residual)

     4. $e^{2h} \leftarrow $ V-CYCLE$(\nu_1, \nu_2, 0, r^{2h})$ (solve $\mathscr{L}_{2h} e^{2h} = r^{2h}$)

     5. $e^h \leftarrow I_{2h}^h e^{2h}$ (compute fine grid error)

     6. $v^h \leftarrow v^h + e^h$ (correct)

     7. $v^h \leftarrow \mathscr{S}^{\nu_2}(v^h, g^h)$ (post-smooth)

- end

In the above V-cycle we described how to interpolate grid functions be-

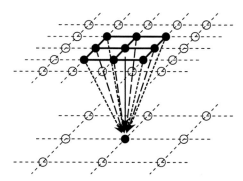

Figure 4.13: Full-weighting restriction

tween finer and coarser grids by restriction with I_h^{2h} and prolongation with I_{2h}^h. Nevertheless, we have not specified how to choose the coarse grid operator $\mathcal{L}_{2h}$ from the fine grid operator $\mathcal{L}_h$. There are two approaches for determining $\mathcal{L}_{2h}$ from $\mathcal{L}_h$.

Direct coarsening. We choose $\mathcal{L}_{2h}$ as the direct analogon of $\mathcal{L}_h$ on the coarse grid Ω_{2h}. That is, we apply the definition of $\mathcal{L}_h$ for the doubled grid spacing $2h$.

Galerkin coarsening. Let $\hat{I}_{2h}^h$ and $\hat{I}_h^{2h}$ be a prolongation and restriction operator, respectively. The so-called *Galerkin coarse grid operator* is defined as

$$\mathcal{L}_{2h} = \hat{I}_h^{2h} \mathcal{L}_h \hat{I}_{2h}^h.$$

Note that $\hat{I}_{2h}^h$ and $\hat{I}_h^{2h}$ must not necessarily coincide with the interpolation operators I_{2h}^h and I_h^{2h} used for grid functions.

The Galerkin coarsening approach is computationally more involved than direct coarsening. In particular, for operators with spatially varying coefficients and so-called algebraic multigrid methods, the Galerkin coarsening plays an important role. Here, however, $\mathcal{L}_h$ is spatially invariant with constant coefficients and we will apply computationally less involved direct coarsening to compute $\mathcal{L}_{2h}$.

Next, we give standard interpolation and smoothing operators to solve (4.30) in two dimensions. To this end let $\Omega = (0,1)^2$ be the unit square and

$$\Omega_h := \{x_{ij} = (i\,h_1, j\,h_2)^\top \ : \ i = 0, 1, \ldots, N_1, \ j = 0, 1, \ldots, N_2\}$$

a grid-point symmetric discretization with $h_\ell = 1/N_\ell$ and N_ℓ a power of two.

For restricting grid functions, we use the so-called *full-weighting restriction*. Let $u^h \in \mathcal{G}(\Omega_h; \mathbb{R}^2)$ be a grid function. The full-weighting restriction

$I_h^{2h} u^h = u^{2h} \in \mathcal{G}(\Omega_{2h}; \mathbb{R}^2)$ of u^h is given by

$$
\begin{aligned}
u_{i,j}^{2h} = &\tfrac{1}{2} u_{2i,2j}^h + \\
&\tfrac{1}{4}(u_{2i+1,2j}^h + u_{2i-1,2j}^h + u_{2i,2j+1}^h + u_{2i,2j-1}^h) + \\
&\tfrac{1}{16}(u_{2i+1,2j+1}^h + u_{2i-1,2j+1}^h + u_{2i+1,2j-1}^h + u_{2i-1,2j-1}^h),
\end{aligned}
$$

for $i = 0, 1, \ldots, N_1/2$ and $j = 0, 1, \ldots, N_2/2$. For the restriction of points with index $i \in \{0, N_1\}$ or $j \in \{0, N_2\}$ we need values from outside of Ω_h. They must be amended according to the imposed boundary conditions (cf. section 4.4.3). Figure 4.13 illustrates the full-weighting restriction of a single point.

To prolongate grid functions we use common bilinear interpolation. Let $u^{2h} \in \mathcal{G}(\Omega_{2h}; \mathbb{R}^2)$ and $I_{2h}^h u^{2h} = u^h \in \mathcal{G}(\Omega_h; \mathbb{R}^2)$ be its prolongated version obtained by bilinear interpolation. Then we have four classes of points for u^h. This is illustrated by Figure 4.14 (a)-(d) and they are given by

$$
\begin{aligned}
u_{2i,2j}^h &= u_{i,j}^{2h} & \text{(a)}, \\
u_{2i+1,2j}^h &= \tfrac{1}{2}(u_{i,j}^{2h} + u_{i+1,j}^{2h}) & \text{(b)}, \\
u_{2i,2j+1}^h &= \tfrac{1}{2}(u_{i,j}^{2h} + u_{i,j+1}^{2h}) & \text{(c)}, \\
u_{2i+1,2j+1}^h &= \tfrac{1}{4}(u_{i,j}^{2h} + u_{i,j+1}^{2h} + u_{i+1,j}^{2h} + u_{i+1,j+1}^{2h}) & \text{(d)}.
\end{aligned}
$$

for $i = 0, 1, \ldots, N_1/2$ and $j = 0, 1, \ldots, N_2/2$.

Finally we need a smoother. Therefore, we apply the so-called *collective ω-Jacobi relaxation*. In general, $\mathscr{L}_h := \mathrm{id} + \tau \alpha \mathscr{A}_h$ reads for two dimensions as

$$
\mathscr{L}_h = \begin{pmatrix} S_h^{11} & S_h^{12} \\ S_h^{21} & S_h^{22} \end{pmatrix} \quad \text{with stencils} \quad S_h^{\mu\nu} = \left[s_{ij}^{\mu\nu} \right]_h, \; i, j \in \mathbb{Z}, \; \mu, \nu = 1, 2.
$$

Let $u^h = (u_1^h, u_2^h)$, $g^h = (g_1^h, g_2^h)$, and $v^h = (v_1^h, v_2^h) \in \mathcal{G}(\Omega_h; \mathbb{R}^2)$ be grid functions with components $u_\ell^h, g_\ell^h, v_\ell^h \in \mathcal{G}(\Omega_h; \mathbb{R})$, $\ell = 1, 2$. Then the collective ω-Jacobi relaxation is defined as

$$
v^h = \mathscr{S}(u^h, g^h)
$$

where

$$
\begin{pmatrix} v_{1,ij}^h \\ v_{2,ij}^h \end{pmatrix} = \begin{pmatrix} u_{1,ij}^h \\ u_{2,ij}^h \end{pmatrix} + \omega \begin{pmatrix} s_0^{11} & s_0^{12} \\ s_0^{21} & s_0^{22} \end{pmatrix}^{-1} \begin{pmatrix} g_{1,ij}^h - (S_h^{11} u_{1,ij}^h + S_h^{12} u_{2,ij}^h) \\ g_{2,ij}^h - (S_h^{21} u_{1,ij}^h + S_h^{22} u_{2,ij}^h), \end{pmatrix} \quad (4.31)
$$

for $i = 0, 1, \ldots, N_1$, $j = 0, 1, \ldots, N_2$ and fixed $\omega \in \mathbb{R}$. Thus, we have to solve a small 2×2 system at each grid point. Fortunately, in our cases the off-diagonal entries s_0^{12} and s_0^{21} are zeros. Recall that the dimensions for $\mathscr{A}^{\mathrm{DIFF}}$ and $\mathscr{A}^{\mathrm{CURV}}$ decouples and therefore $\mathrm{id} + \tau \alpha \mathscr{A}_h$ takes the form

$$
\mathscr{L}_h = \mathrm{id} + \tau \alpha \mathscr{A}_h = \begin{pmatrix} S_h^{11} & \\ & S_h^{22} \end{pmatrix}
$$

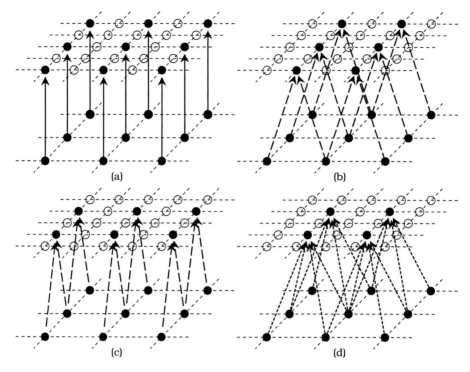

Figure 4.14: Linear prolongation

where, additionally $S_h^{11} = S_h^{22}$. The stencil for elastic operator $\mathscr{A}_h^{\text{ELAS}}$ in two dimensions was given by

$$
\mathscr{A}_h^{\text{ELAS}} = - \left(
\begin{bmatrix}
 & \frac{\mu}{h_2^2} & \\
\frac{2\mu+\lambda}{h_1^2} & -2\left(\frac{2\mu+\lambda}{h_1^2}+\frac{\mu}{h_2^2}\right) & \frac{2\mu+\lambda}{h_1^2} \\
 & \frac{\mu}{h_2^2} &
\end{bmatrix}_h
\begin{bmatrix}
-\frac{\mu+\lambda}{4h_1h_2} & & \frac{\mu+\lambda}{4h_1h_2} \\
 & 0 & \\
\frac{\mu+\lambda}{4h_1h_2} & & -\frac{\mu+\lambda}{4h_1h_2}
\end{bmatrix}_h \\[2em]
\begin{bmatrix}
-\frac{\mu+\lambda}{4h_1h_2} & & \frac{\mu+\lambda}{4h_1h_2} \\
 & 0 & \\
\frac{\mu+\lambda}{4h_1h_2} & & -\frac{\mu+\lambda}{4h_1h_2}
\end{bmatrix}_h
\begin{bmatrix}
 & \frac{2\mu+\lambda}{h_2^2} & \\
\frac{\mu}{h_1^2} & -2\left(\frac{\mu}{h_1^2}+\frac{2\mu+\lambda}{h_2^2}\right) & \frac{\mu}{h_1^2} \\
 & \frac{2\mu+\lambda}{h_2^2} &
\end{bmatrix}_h
\right).
$$

Thus, we obtain for $\mathscr{L}_h = \text{id} + \tau\alpha\mathscr{A}_h^{\text{ELAS}} = (S_h^{\mu\nu})_{\mu,\nu=1}^2$ the stencils

$$
S_h^{11} = \begin{bmatrix}
 & -\frac{\mu}{h_2^2} & \\
-\frac{2\mu+\lambda}{h_1^2} & 1+2\tau\alpha\left(\frac{2\mu+\lambda}{h_1^2}+\frac{\mu}{h_2^2}\right) & -\frac{2\mu+\lambda}{h_1^2} \\
 & \frac{\mu}{h_2^2} &
\end{bmatrix}_h, \quad S_h^{22} = (S_h^{11})^\top,
$$

and

$$
S_h^{12} = S_h^{21} = \begin{bmatrix}
\frac{\mu+\lambda}{4h_1h_2} & & -\frac{\mu+\lambda}{4h_1h_2} \\
 & 0 & \\
-\frac{\mu+\lambda}{4h_1h_2} & & \frac{\mu+\lambda}{4h_1h_2}
\end{bmatrix}_h.
$$

ν_1, ν_2	$h_1 = h_2 = \frac{1}{32}$		$h_1 = h_2 = \frac{1}{64}$		$h_1 = h_2 = \frac{1}{128}$		$h_1 = h_2 = \frac{1}{256}$	
	$\overline{\kappa}^{(20)}$	$\kappa^{(20)}$	$\overline{\kappa}^{(20)}$	$\kappa^{(20)}$	$\overline{\kappa}^{(20)}$	$\kappa^{(20)}$	$\overline{\kappa}^{(20)}$	$\kappa^{(20)}$
$1,1$	0.5849	0.5921	0.5943	0.5986	0.5906	0.5969	0.5941	0.5986
$2,1$	0.4583	0.4652	0.4646	0.4671	0.4638	0.4711	0.4649	0.4694
$2,2$	0.3599	0.3657	0.3633	0.3664	0.3630	0.3701	0.3634	0.3687
$3,1$	0.3627	0.3659	0.3659	0.3680	0.3676	0.3724	0.3682	0.3720

Table 4.5: Empirical convergence rates of V-CYCLE($\nu_1, \nu_2, v^{h,20}, 0$) for solving $(\text{id} + \tau\alpha\mathscr{A}_h^{\text{ELAS}})u^h = 0$ with $\tau = \alpha = \mu = 1$, $\lambda = 0$, $m_0 = 5$ and randomly chosen initial guess $v^{h,0} \in \mathcal{G}(\Omega_h; [0, 1/h_1]^2)$

Therefore (4.31) simplifies to

$$\begin{pmatrix} v_{1,ij}^h \\ v_{2,ij}^h \end{pmatrix} = \begin{pmatrix} u_{1,ij}^h \\ u_{2,ij}^h \end{pmatrix} + \begin{pmatrix} \frac{\omega}{s_0^{11}} \left(g_{1,ij}^h - (S_h^{11} u_{1,ij}^h + S_h^{12} u_{2,ij}^h) \right) \\ \frac{\omega}{s_0^{22}} \left(g_{2,ij}^h - (S_h^{21} u_{1,ij}^h + S_h^{22} u_{2,ij}^h) \right) \end{pmatrix}.$$

Figure 4.15 shows an example for an implementation in Matlab of a multigrid V-cycle for solving $\text{id} + \tau\alpha\mathscr{A}_h^{\text{ELAS}}$ with Dirichlet boundary conditions using full-weighting restriction, linear interpolation, and collective ω-Jacobi relaxation. Note that the Dirichlet boundary reads $u_{ij}^h = 0$ if $i \in \{0, N_1\}$ or $j \in \{0, N_2\}$. To this end, we only have to process the points $i = 1, 2 \ldots, N_1 - 1$ and $j = 1, 2 \ldots, N_2 - 1$.

However, multigrid is an iterative method and in general a single V-cycle will not produce an exact solution. Therefore, the interesting property of a multigrid method is the speed of its convergence to the solution. A practical way to measure multigrid convergence rates is to measure the decay of the norm of the residual. To this end, let $v^{h,0}$ be an initial guess, g^h a given right-hand-side, $v^{h,m} = \text{V-CYCLE}(\nu_1, \nu_2, v^{h,m-1}, g^h)$ the solution after applying m subsequent V-cycles, and $r^{h,m} = g^h - \mathscr{L}_h v^{h,m}$ the remaining residual. The empirical convergence rate

$$\kappa^{(m)} := \frac{\|r^{h,m}\|_\infty}{\|r^{h,m-1}\|_\infty}$$

gives the reduction rate for the residual after m iterations, where $\|u^h\|_\infty := \max_{i,j} |u_{ij}^h|$. Furthermore, a common quantity is the algebraic average

$$\overline{\kappa}^{(m)} := \sqrt[m-m_0]{\kappa^{(m)} \kappa^{(m-1)} \cdots \kappa^{(m_0)}} = \sqrt[m-m_0]{\frac{\|r^{h,m}\|_\infty}{\|r^{h,m_0}\|_\infty}},$$

with $0 \leq m_0 < m$. Often the residual decreases strongly in the first few iterations that does not reflect the asymptotic behavior. Therefore, the

```
function v=restrict(u)
G=1/16*[1 2 1;2 4 2;1 2 1];
tmp=conv2(u,G,'same'); v=tmp(2:2:end-1,2:2:end-1);

function v=interpolate(u)
tmp=zeros(size(u)+2);
tmp(2:end-1,2:end-1)=u; tmp=interp2(tmp,1); v=tmp(2:end-1,2:end-1);

function A=getElasticStencil(h,τ,α,μ,λ)
d11=1/h(1)^2        * [0  1  0 ; 0 -2  0 ;  0  1  0];
d22=1/h(2)^2        * [0  0  0 ; 1 -2  1 ;  0  0  0];
d12=1/(4*h(1)*h(2)) * [1  0 -1 ; 0  0  0 ; -1  0  1];
d21=1/(4*h(1)*h(2)) * [1  0 -1 ; 0  0  0 ; -1  0  1];
I   =                 [0  0  0 ; 0  1  0 ;  0  0  0];
A={ I-τ*α*((2*μ+λ)*d11+μ*d22) ,   -τ*α*(μ+λ)*d12
     -τ*α*(μ+λ)*d21            , I-τ*α*(μ*d11+(2*μ+λ)*d22) };

function [Au1,Au2]=evaluate(τ,α,μ,λ,u1,u2)
A  =getElasticStencil(1./(size(u1)+1),τ,α,μ,λ);
Au1=conv2(u1,A{1,1},'same')+conv2(u2,A{1,2},'same');
Au2=conv2(u1,A{2,1},'same')+conv2(u2,A{2,2},'same');

function [u1,u2]=relax(NREL,ω,τ,α,μ,λ,u1,u2,f1,f2)
A=getElasticStencil(1./(size(u1)+1),τ,α,μ,λ);
for ν=1:NREL
  [Au1,Au2]=evaluate(τ,α,μ,λ,u1,u2);
  u1=u1+ω/A{1,1}(2,2)*(f1-Au1);
  u2=u2+ω/A{2,2}(2,2)*(f2-Au2);
end;

function [u1,u2]=vcycle(NRELD,NRELU,NSOLVE,ω,τ,α,μ,λ,u1,u2,g1,g2)
if (size(u1,1)==1) | (size(u1,2)==1)
  [u1,u2]=relax(NSOLVE,1.0,τ,α,μ,λ,u1,u2,g1,g2);
else
  [u1,u2]=relax(NRELD,ω,τ,α,μ,λ,u1,u2,g1,g2);
  [Au1,Au2]=evaluate(τ,α,μ,λ,u1,u2);
  r1=restrict(g1-Au1); r2=restrict(g2-Au2);
  e1=zeros(size(r1));  e2=zeros(size(r2));
  [e1,e2]=vcycle(NRELD,NRELU,NSOLVE,ω,τ,α,μ,λ,e1,e2,r1,r2);
  u1=u1+interpolate(e1); u2=u2+interpolate(e2);
  [u1,u2]=relax(NRELU,ω,τ,α,μ,λ,u1,u2,g1,g2);
end;
```

Figure 4.15: Matlab code example for solving $(I + \tau\alpha\mathscr{A}_h^{\text{ELAS}})u^h = g^h$ in two dimensions with Dirichlet boundary conditions

first m_0 iterations are left out. Common values for m_0 are in the range 2-5 [59]. Table 4.5 lists empirical convergence rates after 20 iterations computed with the code example given in Figure 4.15.

4.6 Numerical Examples

Next we present two numerical examples to demonstrate the behavior and performance of the presented methods. The first example has academic nature and illustrates the differences between the Diffusive, Elastic, and Curvature smoother. The second example shows the alignment of some real-life CT and PET slices.

For all experiments the time steps τ_n were not changed during the iteration and set to an initially chosen fixed value τ. All three parameters α, σ, and τ were chosen manually according to compute "good" alignments while keeping the transformations smooth without any cracks or foldings. In particular to produce a solution that is one-to-one.

For the Diffusive and Elastic registration Dirichlet boundary conditions were imposed and for the experiments with the curvature smoother Neumann boundary conditions were used. Furthermore, the images were represented as cubic B-splines and all derivatives were computed exactly.

A general difficulty of the PDE based approach is the measurement of convergence to minimizer of the joint functional. We are computing a numerical solution that stems from an optimization problem. Therefore, we discretized the PDE but not the optimization problem. Thus, there is no (discrete) objective function behind which can be evaluated. Nevertheless, to show the PDE based registration approach increases the mutual information of the images we estimate the MI of discrete versions of the reference and deformed template using the joint histogram, i.e., we bin the intensities (cf. section 2.6.1 and section 4.5.1).

Furthermore, for the discrete versions of the images we have shown the mutual information is bounded by the entropies of the images (cf. Theorem 2.33). We use this upper bound to compute a normalized distance between the images. To this end, we define the *relative distance* of two images R and T as

$$relative\ distance\ := \frac{H[R] - \mathrm{MI}[R, T]}{H[R]}. \tag{4.32}$$

Here, we estimate the mutual information and the entropy from histograms, such that $\mathrm{MI}[R, T] \leq H[R]$ and hence the relative distance takes only values between 0 and 1.

4.6.1 Example I: Circles to Squares

The first example is the registration of concentric circles to concentric squares where the intensity gradient of the circles goes from black to white and the gradient of the squares conversely is white to black (cf. Figure 4.18).

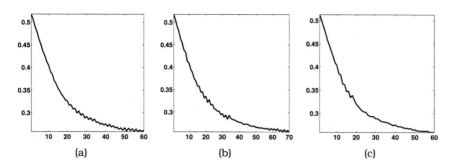

Figure 4.16: Convergence of the PDE based registration for example I. (a) diffusive registration, (b) elastic registration, (c) curvature registration

	#iterations	τ	α	σ	relative distance
Diffusive	60	$5 \cdot 10^{-5}$	2	0.05	0.2612
Elastic ($\mu = 1$, $\lambda = 0$)	70	$5 \cdot 10^{-5}$	2	0.05	0.2559
Curvature	60	$5 \cdot 10^{-5}$	$5 \cdot 10^{-3}$	0.05	0.2591

Table 4.6: Parameters for example I

The purpose of this example is to demonstrate the different levels of smoothness introduced to the transformation by the different smoothers. A transformation that leads to a "perfect" alignment of the images cannot be continuous differentiable everywhere. Due to the change of topology from a circle to a square, the derivatives of a smooth transformation at the edges of a square must tend to infinity the closer the images are aligned. Therefore, the smoothers become large and dominates the influence of the forces and distance, respectively, such that we can study how they effect the resulting transformations.

From Table 4.6 and Figure 4.16 we see that the three methods converge to roughly the same relative distances (≈ 0.3), whereas the alignments are significantly different. Comparing the computed transformations (cf. the second column in Figure 4.18) we observe the several levels of smoothness induced by the three regularizers. The result for the diffusive registration appears to be slightly less smooth than the results for elastic registration, and the curvature registration produced the smoothest transformation of the three. These results are not surprising. Both, the diffusive and elastic smoother penalize first-order derivatives, whereas the curvature smoother is much more restrictive by penalizing second-order derivatives.

Furthermore, the joint histograms are clustered around the diagonal (cf. the third column in Figure 4.18). This reflects that after registration the

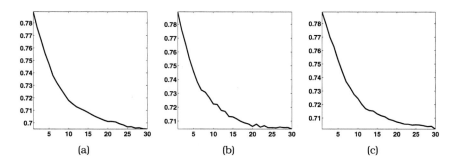

Figure 4.17: Convergence of the PDE method for example II. (a) Diffusive registration, (b) Elastic registration, (c) Curvature registration.

	#iterations	τ	α	σ	relative distance
Diffusive	30	$2 \cdot 10^{-4}$	1	0.1	0.6951
Elastic ($\mu = 1$, $\lambda = 0$)	30	$5 \cdot 10^{-4}$	1	0.1	0.7041
Curvature	30	$3 \cdot 10^{-4}$	10^{-3}	0.1	0.7018

Table 4.7: Parameters for example II

transformed template is approximately an inverted version of the reference image.

4.6.2 Example II: 2D PET-CT Registration

In this example we register 2D slices taken from two real-life 3D CT and PET data sets. The 2D images are coronal slices of the chest showing the lungs and the heart. The CT image shown in Figure 4.19(a) is taken as the reference and shows the patient during an inhalation phase, i.e., the lungs are inflated. Therefore, the PET image shown in Figure 4.19(b) acts as as template. In contrast to the CT data, the PET image is taken from the patient after an exhalation such that the lungs are deflated.

We observe the same behavior as in the first example. Here, the methods converge to roughly the same relative distance of ≈ 0.7, cf. Table 4.7 and Figure 4.17. The result after diffusive registration is at least smooth and the curvature registration produced the smoothest transformation (see Figure 4.19).

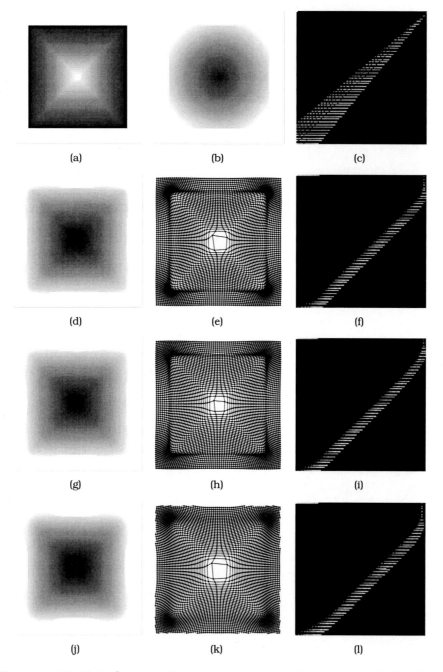

Figure 4.18: (a) Reference, (b) template, (c) joint histogram; (d)–(f) diffusive registration, (g)–(i) elastic registration, (j)–(l) curvature registration; (d), (g), (j) deformed templates; (e), (h), (k) inverse transforms applied to a regular grid; (f), (i), (l) joint histograms of reference and deformed templates

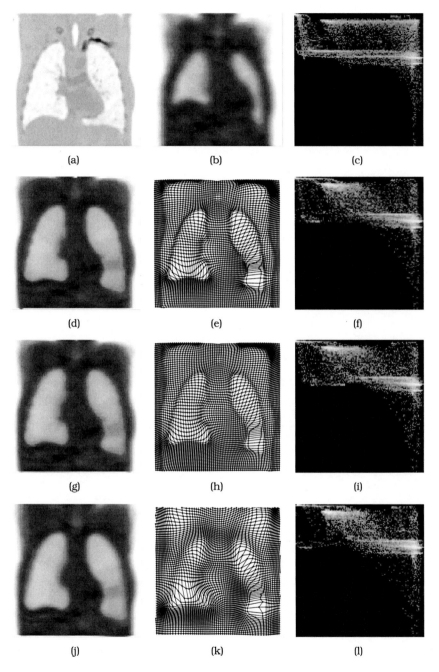

Figure 4.19: (a) Reference, (b) template, (c) joint histogram; (d)–(f) diffusive registration, (g)–(i) elastic registration, (j)–(l) curvature registration; (d), (g), (j) deformed templates; (e), (h), (k) inverse transforms applied to a regular grid; (f), (i), (l) joint histograms of reference and deformed templates

Chapter 5

Optimization Based Methods

In the previous chapter we considered PDE based methods to minimize the joint functional

$$\mathcal{J}[u] = \mathcal{D}[u] + \alpha \mathcal{S}[u]$$

for the registration of images. Therefore, we first optimized by deriving a necessary condition, namely the Euler-Lagrange equations and in a second step we discretized the resulting equations. Therefore, we also call the PDE based approach *optimize-discretize* approach.

Conversely, here we firstly discretize the joint functional and subsequently apply a standard optimization method to find a minimizer of the discretized functional. Therefore, we refer to this approach as *discretize-optimize* approach.

In contrast to PDE based methods, here we view image registration purely as an minimization problem and give up the interpretation of computing continuous motion. We are only interested in finding a minimizer as fast as possible. Therefore, we welcome large intermediate steps as long as they decrease the objective function and yield quick convergence of the method.

5.1 Discretization of the Joint Functional

In this section we establish a discrete version of the joint functional and the distance measure $\mathcal{D}$ and smoother $\mathcal{S}$, respectively.

Let $\Omega \subset \mathbb{R}^d$ be a domain, Ω_h an equispaced discretization of Ω, i.e.,

$$\Omega_h = \{x_k = a + k \odot h \ : \ k = (k_1, \dots, k_d), \ k_\ell = 1, \dots, N_\ell, \ \ell = 1, \dots, d\}$$

and $u^h = (u(x_k))_{x_k \in \Omega_h} \in \mathbb{R}^{N_1 \times \dots \times N_d \times d}$ with $u_k^h := u(x_k) \in \mathbb{R}^d$ the related grid function to $u : \mathbb{R}^d \to \mathbb{R}^d$. For the discretization of the joint functional we consider functions

$$D : \mathbb{R}^{Nd} \to \mathbb{R} \quad \text{and} \quad S : \mathbb{R}^{Nd} \to \mathbb{R}, \qquad N := \#\Omega_h$$

which approximate the distance measure $\mathcal{D}$ and the smoother $\mathcal{S}$, i.e.,

$$D(\vec{u}^h) \approx \mathcal{D}[u] \quad \text{and} \quad S(\vec{u}^h) \approx \mathcal{S}[u]$$

such that

$$J(\vec{u}^h) := D(\vec{u}^h) + \alpha S(\vec{u}^h) \quad \approx \quad \mathcal{J}[u] = \mathcal{D}[u] + \alpha \mathcal{S}[u].$$

Thereby $\vec{u}^h$ denotes the related column vector to u^h obtained by the vec-operation introduced in the previous chapter, cf. (4.24).

Analogous to u and related u^h, $\vec{u}^h$, in the following we consider grid functions and vectors of the images, e.g., for the reference image $R \in \text{Img}(\Omega)$ we have

$$R^h = (R(x_k))_{x_k \in \Omega_h} \in \mathbb{R}^{N_1 \times \dots \times N_d} \quad \text{with} \quad R_k^h := R(x_k) \in \mathbb{R} \text{ and } \vec{R}^h \in \mathbb{R}^N.$$

Furthermore, for the deformable temple image $T \in \text{Img}(\Omega)$ we define T^h as a function of the displacement u^h, i.e., we set

$$T^h : \mathbb{R}^{N_1 \times \dots \times N_d \times d} \rightarrow \mathbb{R}^{N_1 \times \dots \times N_d}, \quad u^h \mapsto T^h(u^h) := \left(T(x_k - u_k^h)\right)_{k\,:\,x_k \in \Omega_h}$$

where $T_k^h(u^h) = T(x_k - u_k^h) \in \mathbb{R}$ and additionally

$$\vec{T}^h : \mathbb{R}^{Nd} \rightarrow \mathbb{R}^N, \quad \vec{u}^h \mapsto \vec{T}^h(\vec{u}^h) := \text{vec}(T^h(u^h)).$$

Later on, we also have need for the Jacobian matrix $D\vec{T}^h(\vec{u}^h) \in \mathbb{R}^{N \times Nd}$ of $\vec{T}^h$ at $\vec{u}^h$. It is given by

$$D\vec{T}^h(\vec{u}^h) = \left(\text{diag}\left(\partial_1 T(x_k - u_k^h)\right)_{k\,:\,x_k \in \Omega_h}, \dots, \text{diag}\left(\partial_d T(x_k - u_k^h)\right)_{k\,:\,x_k \in \Omega_h} \right).$$

Next, we give a discrete version of the MI distance measure.

5.1.1 Discretization of the Distance Measure

For the discretization of the distance measure

$$\mathcal{D}[u] = \mathcal{D}^{\text{MI}}[u] = -\text{MI}[R, T \circ (\text{id} - u)]$$

we proceed in two steps. First, we introduce a binning of gray-values and discretize $\text{MI}[R, T]$ as a function

$$\text{MI} : \mathbb{R}^{M^2} \rightarrow \mathbb{R}, \quad \vec{p} \mapsto \text{MI}(\vec{p}),$$

where M is the number of bins for the gray-values, of a discrete joint density, i.e., a matrix $p \in \mathbb{R}_+^{M \times M}$ which sums to one. Second, we define the joint density as a function of images, i.e.,

$$p : \mathbb{R}^N \times \mathbb{R}^N \rightarrow \mathbb{R}^{M \times M}, \quad (\vec{R}^h, \vec{T}^h) \mapsto p(\vec{R}^h, \vec{T}^h).$$

Finally we set the discrete distance measure to

$$D(\vec{u}^h) := -\text{MI}(\vec{p}(\vec{R}^h, \vec{T}^h(\vec{u}^h))). \tag{5.1}$$

Discretizing Mutual Information

For the definition of a discrete version of

$$\mathrm{MI}[R,T] = \int_{\mathbb{R}\times\mathbb{R}} p_{RT}(r,t) \log \frac{p_{RT}(r,t)}{p_R(r)p_T(t)} \, \mathrm{d}(r,t)$$

we introduce a binning. We proceed analogously to section 2.6.1 on the relation of discrete and absolutely continuous images. Defining intervals $I_{\beta,j} := (j\beta - \frac{\beta}{2}, j\beta + \frac{\beta}{2}]$ for a bin-size $\beta > 0$ and $j \in \mathbb{Z}$ we have from Theorem 2.34

$$\sum_{k,\ell\in\mathbb{Z}} P_{RT}(I_{\beta,k} \times I_{\beta,\ell}) \log \frac{P_{RT}(I_{\beta,k} \times I_{\beta,\ell})}{P_R(I_{\beta,k})\, P_T(I_{\beta,\ell})} \quad \rightarrow \quad \mathrm{MI}[R,T] \quad \text{as } \beta \to 0$$

if the joint density p_{RT} is Riemann integrable. We use this result to develop a function that approximates $\mathrm{MI}[R,T]$. The basic idea is to choose a finite value for the bin-size β and establish an approximation for the probabilities $P_{RT}(I_{\beta,k} \times I_{\beta,\ell})$.

By definition, images are bounded from below and above, such that $P_{RT}(I_{\beta,k} \times I_{\beta,\ell}) = P_R(I_{\beta,k})P_T(I_{\beta,\ell}) = 0$ for k and ℓ small and large enough, respectively. Without loss of generality, in the following we assume the range of both R and T is covered by $\bigcup_{j=1}^{M} I_{\beta,j}$. Defining the matrix

$$p := (P_{RT}(I_{\beta,k} \times I_{\beta,\ell}))_{k,\ell=1}^{M,M} \in \mathbb{R}^{M\times M}$$

we have $p_{k\ell} = P_{RT}(I_{\beta,k} \times I_{\beta,\ell})$ and from Lemma 2.12 follows

$$P_R(I_{\beta,k}) = P_{RT}(I_{\beta,k} \times \mathbb{R}) = \sum_{\ell\in\mathbb{Z}}^{M} P_{RT}(I_{\beta,k} \times I_{\beta,\ell}) = \sum_{\ell=1}^{M} P_{RT}(I_{\beta,k} \times I_{\beta,\ell}) = \sum_{\ell=1}^{M} p_{k\ell}$$

and analogous $P_R(I_{\beta,\ell}) = \sum_{k=1}^{M} p_{k\ell}$. Hence,

$$\sum_{k,\ell=1}^{M} P_{RT}(I_{\beta,k} \times I_{\beta,\ell}) \log \frac{P_{RT}(I_{\beta,k} \times I_{\beta,\ell})}{P_R(I_{\beta,k})\, P_T(I_{\beta,\ell})}$$

$$= \sum_{k,\ell=1}^{M} p_{k\ell} \log \frac{p_{k\ell}}{\left(\sum_{\ell=1}^{M} p_{k\ell}\right)\left(\sum_{k=1}^{M} p_{k\ell}\right)}$$

$$= \sum_{k,\ell=1}^{M} p_{k\ell} \log p_{k\ell} - \sum_{k=1}^{M}\left(\sum_{\ell=1}^{M} p_{k\ell}\right)\log\left(\sum_{\ell=1}^{M} p_{k\ell}\right) - \sum_{\ell=1}^{M}\left(\sum_{k=1}^{M} p_{k\ell}\right)\log\left(\sum_{k=1}^{M} p_{k\ell}\right).$$

We come to the computation of the matrix p and an approximation with the required summation properties soon. For the definition of the MI function we define the entropy function

$$H : \mathbb{R}_+^n \to \mathbb{R}, \quad q \mapsto H(q) := -\sum_{j=1}^{n} q_j \log(q_j + \varepsilon) \tag{5.2}$$

where we add a small $\varepsilon > 0$, e.g., $\varepsilon = 10^{-8}$ to ensure differentiability. Furthermore, with $e := (1, 1, \ldots, 1)^\top \in \mathbb{R}^M$ the one-vector we have

$$\left(\sum_{k=1}^{M} p_{k\ell} \right)_{\ell=1}^{M} = \left(e^\top p \right)^\top = S_1 \vec{p} \quad \text{with } S_1 := I \otimes e^\top \in \mathbb{R}^{M \times M^2}$$

and

$$\left(\sum_{\ell=1}^{M} p_{k\ell} \right)_{k=1}^{M} = p \cdot e = S_2 \vec{p} \quad \text{with } S_2 := e^\top \otimes I \in \mathbb{R}^{M \times M^2}$$

Finally, we define the MI function $\mathrm{MI} : \mathbb{R}^{M^2} \to \mathbb{R}$ as

$$\begin{aligned}
\mathrm{MI}(\vec{p}) &:= H(S_1 \vec{p}) + H(S_2 \vec{p}) - H(\vec{p}) \\
&= \sum_{k,\ell=1}^{M} p_{k\ell} \log \frac{p_{k\ell} + \varepsilon}{\left(\sum_{\kappa=1}^{M} p_{\kappa\ell} + \varepsilon \right) \left(\sum_{\lambda=1}^{M} p_{k\lambda} + \varepsilon \right)}.
\end{aligned} \tag{5.3}$$

In the following we also have need for the gradient $\nabla \mathrm{MI}$. Applying the chain rule we obtain

$$\nabla \mathrm{MI}(\vec{p}) = S_1^\top \nabla H(S_1 \vec{p}) + S_2^\top \nabla H(S_2 \vec{p}) - \nabla H(\vec{p}) \tag{5.4}$$

where

$$\nabla H(q) = - \left(\frac{q_1}{q_1 + \varepsilon} + \log(q_1 + \varepsilon) , \; \ldots \; , \; \frac{q_n}{q_n + \varepsilon} + \log(q_n + \varepsilon) \right)^\top. \tag{5.5}$$

The MI function is quite similar to the expression for the mutual information of discrete images (cf §2.6). The only difference is that we add an ε to ensure differentiability. However, as the mutual information of discrete images, our so-defined MI function is bounded. This is an important property that is crucial for the existence of a minimizer.

Lemma 5.1

Let $p = (p_{k\ell}) \in \mathbb{R}^{M \times M}$ having non-negative entries $p_{k\ell} \geq 0$ and $\sum_{k,\ell=1}^{M} p_{k\ell} = 1$.

Then

$$1 - (1 + M\varepsilon)^2 \; \leq \; \mathrm{MI}(\vec{p}) \; \leq \; \min\{ H(S_1 \vec{p}), \, H(S_2 \vec{p}) \}.$$

Proof. We start with the lower bound. As in the proof of Theorem 2.18b we make use of the inequality $x \log y = x \log x + y - x$ yielding

$$\text{MI}(\vec{p}) = \sum_{k,\ell=1}^{M} p_{k\ell} \log \frac{p_{k\ell} + \varepsilon}{\left(\sum_{\kappa=1}^{M} p_{\kappa\ell} + \varepsilon\right)\left(\sum_{\lambda=1}^{M} p_{k\lambda} + \varepsilon\right)}$$

$$= \sum_{k,\ell=1}^{M} p_{k\ell} \log(p_{k\ell} + \varepsilon) - \sum_{k,\ell=1}^{M} p_{k\ell} \log\left(\left(\sum_{\kappa=1}^{M} p_{\kappa\ell} + \varepsilon\right)\left(\sum_{\lambda=1}^{M} p_{k\lambda} + \varepsilon\right)\right)$$

$$\geq \sum_{k,\ell=1}^{M} p_{k\ell} \log p_{k\ell} - \sum_{k,\ell=1}^{M} p_{k\ell} \log\left(\left(\sum_{\kappa=1}^{M} p_{\kappa\ell} + \varepsilon\right)\left(\sum_{\lambda=1}^{M} p_{k\lambda} + \varepsilon\right)\right)$$

$$\geq \sum_{k,\ell=1}^{M} p_{k\ell} - \sum_{k,\ell=1}^{M} \left(\sum_{\kappa=1}^{M} p_{\kappa\ell} + \varepsilon\right)\left(\sum_{\lambda=1}^{M} p_{k\lambda} + \varepsilon\right)$$

$$= 1 - \left(\sum_{k,\ell=1}^{M}\left(\sum_{\kappa=1}^{M} p_{\kappa\ell}\right)\left(\sum_{\lambda=1}^{M} p_{k\lambda}\right) + \varepsilon \sum_{k,\ell=1}^{M}\sum_{\kappa=1}^{M} p_{\kappa\ell} + \varepsilon \sum_{k,\ell=1}^{M}\sum_{\lambda=1}^{M} p_{k\lambda} + M^2 \varepsilon^2\right)$$

$$= 1 - (1 + 2M\varepsilon + M^2\varepsilon^2) = 1 - (1 + M\varepsilon)^2.$$

Now we turn to the proof of the upper bound. Since $\sum_{\ell=1}^{M} p_{k\ell} \geq p_{k\ell}$ for all k we have

$$\log \frac{p_{k\ell} + \varepsilon}{\left(\sum_{\kappa=1}^{M} p_{\kappa\ell} + \varepsilon\right)\left(\sum_{\lambda=1}^{M} p_{k\lambda} + \varepsilon\right)} \leq \log \frac{\sum_{\ell=1}^{M} p_{k\ell} + \varepsilon}{\left(\sum_{\kappa=1}^{M} p_{\kappa\ell} + \varepsilon\right)\left(\sum_{\lambda=1}^{M} p_{k\lambda} + \varepsilon\right)}$$

$$= -\log\left(\sum_{\kappa=1}^{M} p_{\kappa\ell} + \varepsilon\right)$$

and therefore $\text{MI}(\vec{p}) \leq H(S_1\vec{p})$. Analogous, $\sum_{k=1}^{M} p_{k\ell} \geq p_{k\ell}$ yields $\text{MI}(\vec{p}) \leq H(S_2\vec{p})$. ∎

In particular the upper bound is important for us. Since we want to mini-mize the negative mutual information, this becomes a lower bound of the discrete distance measure $D(\vec{u}^h) = -\text{MI}(\vec{p}(\vec{R}^h, \vec{T}^h(\vec{u}^h)))$. In the following, we will show how to compute $p(\vec{R}^h, \vec{T}^h)$ such that the sum $\sum_\ell p_{k\ell}(\vec{R}^h, \vec{T}^h)$ only depends on the reference R^h. Therefore, $S_1\vec{p}(\vec{R}^h, \vec{T}^h(\vec{u}^h))$ does not de-pend on the deformed template $\vec{T}^h(\vec{u}^h)$ and in particular is independent from the displacement $\vec{u}^h$. Thus, $H(S_1\vec{p}(\vec{R}^h, \vec{T}^h(\vec{u}^h)))$ is constant during the optimization such that the discrete distance measures is bounded from below by

$$-H(S_1\vec{p}(\vec{R}^h, \vec{T}^h)) \leq D(\vec{u}^h) \quad \text{for all } \vec{u}^h \in \mathbb{R}^{Nd}.$$

The Density Function

Now we work out the computation of a matrix $p \in \mathbb{R}_+^{M \times M}$ with $\sum_{k,\ell} p_{k\ell} = 1$ and

$$p_{k\ell} \approx P_{RT}(I_{\beta,k} \times I_{\beta,\ell}).$$

In principle, here we could approximate the joint distribution by computing the joint histogram, i.e., we set

$$p_{k\ell} = \frac{\#\{j \,:\, R(x_j) \in I_{\beta,k} \,\wedge\, T(x_j) \in I_{\beta,\ell}\}}{N}.$$

The drawback is of this proceeding is that we cannot differentiate w.r.t. R and T. Using the joint histogram as a building block for our objective function results a non-differentiable function. Thus, we are restricted to derivative-free optimization, e.g., we could use the Nelder-Mead method. In general, optimization methods that use gradient information converge much faster than a direct method. To this end, we develop a "smoothed" histogram function which is differentiable. We use a discretized version of the approximating density introduced in chapter 3, i.e.,

$$
\begin{aligned}
p_{RT}^{\sigma}(r,t) &= \frac{1}{|\Omega|} \int_{\Omega} \frac{1}{\sigma^2} K\left(\frac{r - R(x)}{\sigma}\right) K\left(\frac{t - T(x)}{\sigma}\right) \, dx \\
&\approx \frac{h}{\sigma^2 |\Omega|} \sum_{x_j \in \Omega_h} K\left(\frac{r - R(x_j)}{\sigma}\right) K\left(\frac{t - T(x_j)}{\sigma}\right),
\end{aligned}
$$

with $h := \prod_{j=1}^{d} h_j$. Then, defining $r_k = \beta k$ and $t_\ell = \beta \ell$ the centers of $I_{\beta,k}$ $I_{\beta,\ell}$, we obtain the approximation

$$
\begin{aligned}
P_{RT}(I_{\beta,k} \times I_{\beta,\ell}) &\approx \int_{I_{\beta,k} \times I_{\beta,\ell}} p_{RT}^{\sigma}(r,t) \, d(r,t) \\
&\approx \beta^2 p_{RT}^{\sigma}(r_k, t_\ell) \\
&\approx \frac{\beta^2 h}{\sigma^2 |\Omega|} \sum_{x_j \in \Omega_h} K\left(\frac{r_k - R(x_j)}{\sigma}\right) K\left(\frac{t_\ell - T(x_j)}{\sigma}\right) =: p_{k\ell},
\end{aligned}
$$

where we applied the mid-point rule to the integral over $I_{\beta,k} \times I_{\beta,\ell}$ with volume $\mathrm{vol}(I_{\beta,k} \times I_{\beta,\ell}) = \beta^2$. Choosing $\beta = \sigma$ and the grid spacing h such that $h = \frac{|\Omega|}{\#\Omega_h}$ we have

$$
\begin{aligned}
p_{k\ell} &= \frac{\beta^2 h}{\sigma^2 |\Omega|} \sum_{x_j \in \Omega_h} K\left(\frac{r_k - R(x_j)}{\sigma}\right) K\left(\frac{t_\ell - T(x_j)}{\sigma}\right) \\
&= \frac{1}{\#\Omega_h} \sum_{x_j \in \Omega_h} K\left(k - \frac{R(x_j)}{\beta}\right) K\left(\ell - \frac{T(x_j)}{\beta}\right).
\end{aligned}
$$

Furthermore, we want the entries $p_{k\ell}$ summing up to one. To ensure this property, we take advantage from choosing K as a B-spline kernel. Here we choose in particular the cubic B-spline given by

$$
K(x) = \frac{1}{6}
\begin{cases}
(x+2)^3 & \text{if } x \in (-2,-1], \\
4 - 6x^2 - 3x^3 & \text{if } x \in (-1,0], \\
4 - 6x^2 + 3x^3 & \text{if } x \in (0,1], \\
-(x-2)^3 & \text{if } x \in (2,2), \\
0 & \text{else.}
\end{cases}
$$

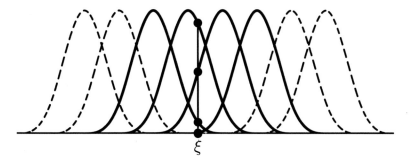

Figure 5.1: Partition of the unity

It is well-known that B-splines form a so-called *partition of the unity* (cf. Figure 5.1), i.e.,

$$\sum_{k \in \mathbb{Z}} K(k + \xi) = 1 \quad \text{for all } \xi \in \mathbb{R}.$$

Thus, summing over the bin-center results

$$\sum_{k=1}^{M}\sum_{\ell=1}^{M} p_{k\ell} = \sum_{k=1}^{M}\sum_{\ell=1}^{M} \left(\frac{1}{\#\Omega_h} \sum_{x_j \in \Omega_h} K\left(k - \frac{R(x_j)}{\beta}\right) K\left(\ell - \frac{T(x_j)}{\beta}\right)\right)$$

$$= \frac{1}{\#\Omega_h} \sum_{x_j \in \Omega_h} \sum_{k=1}^{M} K\left(k - \frac{R(x_j)}{\beta}\right) \sum_{\ell=1}^{M} K\left(\ell - \frac{T(x_j)}{\beta}\right)$$

$$= 1.$$

To this end we define our density function $p : \mathbb{R}^N \times \mathbb{R}^N \to \mathbb{R}^{M \times M}$ by

$$p(\vec{R}^h, \vec{T}^h) := \left(\frac{1}{N} \sum_{j=1}^{N} K\left(k - \frac{\vec{R}_j^h}{\beta}\right) K\left(\ell - \frac{\vec{T}_j^h}{\beta}\right) \right)_{k,\ell=1}^{M,M}$$

such that $p_{k\ell}(\vec{R}^h, \vec{T}^h) \geq 0$, $\sum_{k\ell} p_{k\ell}(\vec{R}^h, \vec{T}^h) = 1$. Furthermore, as mentioned above,

$$\sum_{\ell=1}^{M} p_{k\ell} = \sum_{\ell=1}^{M} \left(\frac{1}{\#\Omega_h} \sum_{x_j \in \Omega_h} K\left(k - \frac{R(x_j)}{\beta}\right) K\left(\ell - \frac{T(x_j)}{\beta}\right)\right)$$

$$= \frac{1}{\#\Omega_h} \sum_{x_j \in \Omega_h} K\left(k - \frac{R(x_j)}{\beta}\right) \underbrace{\left(\sum_{\ell=1}^{M} K\left(\ell - \frac{T(x_j)}{\beta}\right)\right)}_{=1}$$

$$= \frac{1}{\#\Omega_h} \sum_{x_j \in \Omega_h} K\left(k - \frac{R(x_j)}{\beta}\right)$$

shows that $\sum_{\ell} p_{k\ell}(\vec{R}^h, \vec{T}^h)$ is independent from T^h.

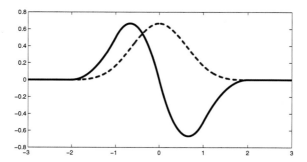

Figure 5.2: Cubic B-spline (dashed line) and its derivative (solid line).

The discrete distance measure (5.1) is defined on the density function as a vector. Therefore, we rearrange the matrix $p(\vec{R}^h, \vec{T}^h)$ as a vector, i.e., we define

$$\vec{p}: \mathbb{R}^N \times \mathbb{R}^N \rightarrow \mathbb{R}^{M^2}, \quad (\vec{R}^h, \vec{T}^h) \mapsto \vec{p}(\vec{R}^h, \vec{T}^h) := \mathrm{vec}(p(\vec{R}^h, \vec{T}^h)).$$

Then, according to the definition of the vec-operator the i-th entry of $\vec{p}(\vec{R}^h, \vec{T}^h)$ is given by

$$\vec{p}_i(\vec{R}^h, \vec{T}^h) = p_{k\ell}(\vec{R}^h, \vec{T}^h) \quad \Leftrightarrow \quad i = k + (\ell - 1)M.$$

Furthermore, for the gradient of the discrete distance measure we need the Jacobian matrix of $\vec{p}$ w.r.t. to $\vec{T}^h$. It is given by

$$\mathrm{D}_{\vec{T}^h}\vec{p}(\vec{R}^h, \vec{T}^h) = \left(\frac{\partial}{\partial \vec{T}_j^h} \vec{p}_i(\vec{R}^h, \vec{T}^h)\right)_{i,j=1}^{M^2,N} \in \mathbb{R}^{M^2 \times N}$$

with entries

$$\frac{\partial}{\partial \vec{T}_j^h} \vec{p}_i(\vec{R}^h, \vec{T}^h) = \frac{\partial}{\partial \vec{T}_j^h} p_{k\ell}(\vec{R}^h, \vec{T}^h) = -\frac{1}{N\beta} K\left(k - \frac{\vec{R}_j^h}{\beta}\right) K'\left(\ell - \frac{\vec{T}_j^h}{\beta}\right),$$

where $i = k + (\ell - 1)M$ and the derivative K' of K given by (cf. Figure 5.2)

$$K'(x) = \frac{1}{6} \begin{cases} 3(x+2)^2 & \text{if } x \in (-2, -1], \\ -12x - 9x^2 & \text{if } x \in (-1, 0], \\ -12x + 9x^2 & \text{if } x \in (0, 1], \\ -3(x-2)^2 & \text{if } x \in (2, 2), \\ 0 & \text{else.} \end{cases}$$

Efficient Computation of the Density Function

An important point for practical computation is that we can evaluate the joint density and its gradient quite efficiently by an $\mathcal{O}(N)$ method. Furthermore, the Jacobian $\mathrm{D}_{\vec{T}^h}\vec{p}(\vec{R}^h, \vec{T}^h)$ is a sparse matrix, i.e., only a few entries are non-zero.

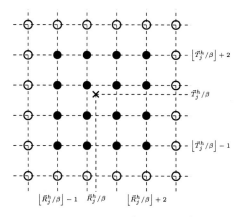

Figure 5.3: Non-zero kernels $K(k - \frac{\vec{R}_j^h}{\beta})K(\ell - \frac{\vec{T}_j^h}{\beta})$ centered at $(k, \ell) \in \mathbb{Z}^2$.

The computational power stems from the compactly supported B-spline kernels. For both K and its derivative K' we have

$$K(x) = K'(x) = 0 \quad \text{if} \quad |x| \geq 2$$

and hence

$$\left. \begin{array}{c} K\left(k - \frac{\vec{R}_j^h}{\beta}\right) bK\left(\ell - \frac{\vec{T}_j^h}{\beta}\right) \\ K\left(k - \frac{\vec{R}_j^h}{\beta}\right) K'\left(\ell - \frac{\vec{T}_j^h}{\beta}\right) \end{array} \right\} \neq 0 \quad \Leftrightarrow \quad \left|k - \frac{\vec{R}_j^h}{\beta}\right| < 2 \wedge \left|\ell - \frac{\vec{T}_j^h}{\beta}\right| < 2.$$

Let us consider for which values of k and ℓ the last equation is valid. For k we have

$$-2 < k - \vec{R}_j^h/\beta < 2 \quad \Leftrightarrow \quad \vec{R}_j^h/\beta - 2 < k < \vec{R}_j^h/\beta + 2$$
$$\Leftrightarrow \quad \left\lfloor \vec{R}_j^h/\beta \right\rfloor - 2 < k < \left\lceil \vec{R}_j^h/\beta \right\rceil + 2$$
$$\Rightarrow \quad \left\lfloor \vec{R}_j^h/\beta \right\rfloor - 1 \leq k \leq \left\lfloor \vec{R}_j^h/\beta \right\rfloor + 2$$

and analogous for ℓ must hold $\left\lfloor \vec{T}_j^h/\beta \right\rfloor - 1 \leq \ell \leq \left\lfloor \vec{T}_j^h/\beta \right\rfloor + 2$, where $\lfloor \cdot \rfloor : \mathbb{R} \to \mathbb{Z}$ and $\lceil \cdot \rceil : \mathbb{R} \to \mathbb{Z}$ denote the common *floor* and *ceiling* functions given by

$$\lfloor x \rfloor := \max\{z \in \mathbb{Z} \ : \ z \leq x\} \quad \text{and} \quad \lceil x \rceil := \min\{z \in \mathbb{Z} \ : \ z \geq x\}.$$

Thus, a single tuple $(\vec{R}_j^h, \vec{T}_j^h)$ contributes to at most 16 kernel functions (cf. Figure 5.3) centered at

$$(k, \ell) \in \left\{ \left\lfloor \vec{R}_j^h/\beta \right\rfloor - 1, \dots, \left\lfloor \vec{R}_j^h/\beta \right\rfloor + 2, \right\} \times \left\{ \left\lfloor \vec{T}_j^h/\beta \right\rfloor - 1, \dots, \left\lfloor \vec{T}_j^h/\beta \right\rfloor + 2, \right\}.$$

Summarizing, we have the following algorithm for the fast computation of $\vec{p}(\vec{R}^h, \vec{T}^h)$ and $D_{\vec{T}^h}\vec{p}(\vec{R}^h, \vec{T}^h)$:

- Initialize $p \in \mathbb{R}^{M^2}$ and $g \in \mathbb{R}^{M^2 \times N}$ with zeros.

- `for` $j = 1, 2, \ldots, N$ `do`

   1. $\varrho \leftarrow \left\lfloor \vec{R}_j^h / \beta \right\rfloor$ and $\vartheta \leftarrow \left\lfloor \vec{T}_j^h / \beta \right\rfloor$,

   2. `for all` $(k, \ell) \in \{\varrho - 1, \ldots, \varrho + 2\} \times \{\vartheta - 1, \ldots, \vartheta + 2\}$ `do`

   $$i \quad \leftarrow k + (\ell - 1)M$$

   $$p_i \quad \leftarrow p_i + \tfrac{1}{N} K\left(k - \tfrac{\vec{R}_j^h}{\beta}\right) K\left(\ell - \tfrac{\vec{T}_j^h}{\beta}\right)$$

   $$g_{i,j} \quad \leftarrow -\tfrac{1}{N\beta} K\left(k - \tfrac{\vec{R}_j^h}{\beta}\right) K'\left(\ell - \tfrac{\vec{T}_j^h}{\beta}\right)$$

   `end`

- `end`

The method requires exactly $16N$ steps and terminates with $\vec{p}(\vec{R}^h, \vec{T}^h) = p$ and $D_{\vec{T}^h}\vec{p}(\vec{R}^h, \vec{T}^h) = g$. Figure 5.4 shows a simple implementation in Matlab.

5.1.2 Discretizing the Smoother

For the discretization of the smoother $\mathcal{S}$ we can hark back to results derived in the previous chapter on PDE based methods. Recalling from chapter 4, the diffusive, curvature, and elastic smoother in general reads

$$\mathcal{S}[u] = \tfrac{1}{2} \langle \mathcal{L}u, \mathcal{L}u \rangle_{L^2(\Omega)}$$

with $\mathcal{L}$ a linear differential operator, cf. equations (4.7), (4.8), (4.11). Furthermore, we computed the Gateâux derivatives of the smoothers which generally take the form

$$d\mathcal{S}[u; v] = \langle \mathcal{A}u, v \rangle_{L^2(\Omega)} \quad + \quad \text{boundary integrals} ,$$

cf. Lemmas 4.4, 4.7, 4.10. Subsequent considerations showed that imposing appropriate boundary conditions make the boundary integrals vanish such that $d\mathcal{S}[u; v] = \langle \mathcal{A}u, v \rangle_{L^2(\Omega)}$. The following Lemma shows that we can rewrite the smoother by its Gateâux derivative.

```
function [p,dp] = computeDensity(R,T,M,beta)
K=inline([ ...
    '((x> -2) & (x<-1)) .* 1/6 .* (x+2).^3            + ' ...
    '((x>=-1) & (x< 0)) .* 1/6 .* (4 - 6*x.^2 - 3*x.^3) + ' ...
    '((x>= 0) & (x< 1)) .* 1/6 .* (4 - 6*x.^2 + 3*x.^3) + ' ...
    '((x>= 1) & (x< 2)) .* 1/6 .* (-(x-2).^3)          ' ...
    ]);

dK=inline([ ...
    '((x> -2) & (x<-1)) .* 1/6 .* (3.*(x+2).^2)     + ' ...
    '((x>=-1) & (x< 0)) .* 1/6 .* (-12*x - 9*x.^2) + ' ...
    '((x>= 0) & (x< 1)) .* 1/6 .* (-12*x + 9*x.^2) + ' ...
    '((x>= 1) & (x< 2)) .* 1/6 .* (-3.*(x-2).^2)     ' ...
    ]);

N  = prod(size(R));
p  = zeros(M^2,1);
dp = sparse([],[],[],M^2,N,16*N);

for j=1:N
    rho     = floor(R(j)/beta);
    theta   = floor(T(j)/beta);
    [k,l]   = ndgrid([(rho-1):(rho+2)],[(theta-1):(theta+2)]);
    k       = reshape(k,16,1);
    l       = reshape(l,16,1);
    i       = k+(l-1)*M;

    p(i)    = p(i) + 1/N * K(k-R(j)/beta) .* K(l-T(j)/beta);
    dp(i,j) = -1/(beta*N)* K(k-R(j)/beta) .* dK(l-T(j)/beta);
end
```

Figure 5.4: Matlab code example for computing the density function ($\mathrm{p} = \vec{p}(\vec{R}^h, \vec{T}^h)$) and its Jacobian matrix ($\mathrm{dp} = \mathrm{D}_{\vec{T}^h}\vec{p}(\vec{R}^h, \vec{T}^h)$)

Lemma 5.2

Let $\Omega \subset \mathbb{R}^d$ be a domain and $\mathscr{L} : C^m(\overline{\Omega}) \to C(\overline{\Omega})$, $\mathscr{A} : C^{2m}(\overline{\Omega}) \to C(\overline{\Omega})$ linear operators. Furthermore, let $S : V \subseteq C^{2m}(\overline{\Omega}) \to \mathbb{R}$ be a functional with

$$S[u] = \tfrac{1}{2} \langle \mathscr{L}u, \mathscr{L}u \rangle_{L^2(\Omega)} \quad \text{and} \quad dS[u; v] = \langle \mathscr{A}u, v \rangle_{L^2(\Omega)}.$$

Then

$$\langle \mathscr{L}u, \mathscr{L}u \rangle_{L^2(\Omega)} = \langle \mathscr{A}u, u \rangle_{L^2(\Omega)} \quad \text{for all } u \in V \quad \text{and} \quad S[u] = \tfrac{1}{2} dS[u; u].$$

Proof.

$$\begin{aligned}
\langle \mathscr{A}u, u \rangle_{L^2(\Omega)} &= dS[u; u] \\
&= \tfrac{d}{d\tau} S[u + \tau u]\Big|_{\tau=0} \\
&= \tfrac{d}{d\tau} \left(\tfrac{1}{2} \langle \mathscr{L}(u + \tau u), \mathscr{L}(u + \tau u) \rangle_{L^2(\Omega)} \right)\Big|_{\tau=0} \\
&= \tfrac{d}{d\tau} \left(\tfrac{1}{2} \langle \mathscr{L}u, \mathscr{L}u \rangle_{L^2(\Omega)} + \tau \langle \mathscr{L}u, \mathscr{L}u \rangle_{L^2(\Omega)} + \tfrac{\tau^2}{2} \langle \mathscr{L}u, \mathscr{L}u \rangle_{L^2(\Omega)} \right)\Big|_{\tau=0} \\
&= \langle \mathscr{L}u, \mathscr{L}u \rangle_{L^2(\Omega)}.
\end{aligned}$$

∎

Based on the above Lemma, we discretize the smoother using the difference approximation derived in section 4.4. Choosing appropriate boundary conditions we have

$$\begin{aligned}
S[u] &= \tfrac{1}{2} \langle \mathscr{A}u, u \rangle_{L^2(\Omega)} \\
&= \tfrac{1}{2} \int_\Omega u(x)^\top \mathscr{A}u(x)\, dx \\
&\approx \tfrac{h}{2} \sum_{x_k \in \Omega} u(x_k)^\top \mathscr{A}_h u(x_k) \\
&= \tfrac{h}{2} (\vec{u}^h)^\top A \vec{u}^h
\end{aligned}$$

with $h = \prod_{j=1}^d h_j$, $\mathscr{A}_h$ a difference operator approximating $\mathscr{A}$ and A the matrix representation of $\mathscr{A}_h$ (cf. sections 4.4.2, 4.4.4). Therefore, we define the smoother function as

$$S(\vec{u}^h) := \tfrac{h}{2} (\vec{u}^h)^\top A \vec{u}^h.$$

with A one of the matrices A^{DIFF}, A^{CURV}, and A^{ELAS} given in section 4.4.4. Furthermore, the gradient ∇S and the Hessian $\nabla^2 S$ of S are given by

$$\nabla S(\vec{u}^h)^\top = h A \vec{u}^h, \quad \text{and} \quad \nabla^2 S(u) = h A.$$

An overview of the derived approximations and functions in this section is shown Table 5.1.

$$\underline{\text{Objective Function \& Building Blocks}}$$

$J : \mathbb{R}^{Nd} \to \mathbb{R},$	$J(\vec{u}^h) = D(\vec{u}^h) + \alpha S(\vec{u}^h)$	(joint function)
$S : \mathbb{R}^{Nd} \to \mathbb{R},$	$S(\vec{u}^h) = \frac{h}{2}(\vec{u}^h)^\top A\vec{u}^h$	(regularizer)
$D : \mathbb{R}^{Nd} \to \mathbb{R},$	$D(\vec{u}^h) = -\text{MI}(\vec{p}(\vec{R}^h, \vec{T}^h(\vec{u}^h)))$	(distance)
$\vec{T}^h : \mathbb{R}^{Nd} \to \mathbb{R}^N,$	$\vec{T}^h(\vec{u}^h) = \text{vec}\left(T(x_k - u_k^h)\right)_{k\,:\,x_k \in \Omega_h}$	(template)
$\text{MI} : \mathbb{R}^{M^2} \to \mathbb{R},$	$\text{MI}(\vec{p}) = H(S_1\vec{p}) + H(S_2\vec{p}) - H(\vec{p})$	(MI)
$H : \mathbb{R}^n \to \mathbb{R},$	$H(q) = -\sum_{j=1}^n q_j \log(q_j + \varepsilon)$	(entropy)

$$\vec{p} : \mathbb{R}^N \times \mathbb{R}^N \to \mathbb{R}^{M^2},$$

$$\vec{p}(\vec{R}^h, \vec{T}^h) = \text{vec}\left(\frac{1}{N}\sum_{j=1}^N K\left(k - \frac{R_j^h}{\beta}\right) K\left(\ell - \frac{T_j^h}{\beta}\right)\right)_{k,\ell=1}^{M,M} \qquad \text{(joint density)}$$

$$\underline{\text{Gradients}}$$

$$\nabla J(\vec{u}^h) = \nabla D(\vec{u}^h) + \alpha \nabla S(\vec{u}^h)$$

$$\nabla S(\vec{u}^h) = hA\vec{u}^h$$

$$\nabla D(\vec{u}^h) = -\text{D}\vec{T}^h(\vec{u}^h)^\top \cdot \text{D}_{\vec{T}^h}\vec{p}(\vec{R}^h, \vec{T}^h(\vec{u}^h))^\top \cdot \nabla\text{MI}(\vec{p}(\vec{R}^h, \vec{T}^h(\vec{u}^h)))$$

$$\nabla\text{MI}(\vec{p}) = \nabla S_1^\top \nabla H(S_1\vec{p}) + S_2^\top \nabla H(S_2\vec{p}) - \nabla H(\vec{p})$$

$$\nabla H(q) = -\left(\frac{q_1}{q_1+\varepsilon} + \log(q_1 + \varepsilon) , \quad \ldots \quad , \frac{q_n}{q_n+\varepsilon} + \log(q_n + \varepsilon)\right)$$

Table 5.1: Overview of the joint function and its building blocks

5.2   Descent Methods

After establishing the joint function $J : \mathbb{R}^{Nd} :\rightarrow \mathbb{R}$ the discretized registration problem reads

$$J(u) = D(u) + \tfrac{\alpha h}{2} u^\top A u \quad \rightarrow \quad \min, \qquad u \in \mathbb{R}^{Nd}.$$

For the computation of a minimizer we apply a descent method. Generally, such a method reads

- Choose an initial guess $u_0 \in \mathbb{R}^{Nd}$

- for $k = 0, 1, 2, \dots$ do

  1. Compute a *descent direction* $d_k \in \mathbb{R}^{Nd}$, i.e., $d_k^\top \nabla J(u_k) < 0$
  2. Choose a step length $\lambda_k > 0$ such that $J(u_k + \lambda_k d_k) < J(u_k)$
  3. $u_{k+1} \leftarrow u_k + \lambda_k d_k$

  end

A very popular descent method is the *steepest descent method*. Therefore, we choose the negative gradient as descent direction, i.e.,

$$d_k = -\nabla J(u_k).$$

This method is easy to implement and provides in general linear convergence to minima. Another important descent method is *Newton's method*. Here, the descent direction is given by

$$d_k = -\nabla^2 J(u_k)^{-1} \nabla J(u_k).$$

It is well-known that Newton's method converges quadratically provided we are close enough to a minima of J.

However, due to its faster convergence Newton's method seems preferable to the steepest descent method. Nevertheless, it is not reasonable for us. Besides, we must solve a linear system in each iteration for Newton's method. First we have to compute the Hessian $\nabla^2 J(u_k) \in \mathbb{R}^{Nd \times Nd}$ with N the number of pixels/voxels. Thus, computing and storing the Hessian involves a huge amount of memory, such that we could only deal with small problems, e.g., for two-dimensional images with $256 \times 256 = 2^{16}$ pixels the Hessian of J has $2^{17} \times 2^{17} = 2^{34}$ entries!

To this end we use a *quasi Newton method*. These methods do not involve the Hessian but provide super-linear convergence. Here, we propose using the prominent *BFGS quasi Newton Method* (Broyden,Fletcher, Goldfarb, Shanno).

5.3 The BFGS-Method

In this section we shortly outline one of the most popular quasi Newton methods - the BFGS method. Generally, a quasi Newton method for finding minimizer of a continuous differentiable function $f : \mathbb{R}^n \to \mathbb{R}$ reads

$$x_{k+1} = x_k - H_k^{-1} \nabla f(x_k)$$

with matrices $H_k \in \mathbb{R}^{n \times n}$ which fulfill the so-called *secant equation* or also called *quasi Newton equation*

$$H_{k+1}(x_{k+1} - x_k) = \nabla f(x_{k+1}) - \nabla f(x_k).$$

An example for matrices fulfilling the secant equation is choosing $H_k = \nabla^2 f(x_k)$ (if f is two-times continuous differentiable) yielding the Newton method. However, we switched to the quasi Newton framework since working with the Hessian $\nabla^2 J$ involves too much storage and computational effort for solving the linear system $\nabla^2 J(x_k) d_k = -\nabla J(x_k)$. The idea of a quasi Newton method is to compute H_{k+1} by a "cheap" update rule from its predecessor H_k and the inverse H_{k+1}^{-1} by updating H_k^{-1}, respectively. Furthermore, to solve our storage problem, instead of building the matrices H_k^{-1} we only keep the information needed for the updates. As we will see, therefore we only have to keep two vectors per iteration.

However, the matrices H_k are not uniquely determined from the secant equation. Therefore, one has to require additional constraints, yielding several quasi Newton methods. Many methods have been developed. Here, we consider the popular BFGS method.

For the BFGS method in addition to the secant equation we require H_k to be symmetric and positive definite. The BFGS update rule is given by (cf. [13, §9.2] or [64, §7])

$$H_{k+1} = H_k - \frac{(H_k s_k)(H_k s_k)^\top}{s_k^\top H_k s_k} + \frac{y_k y_k^\top}{y_k^\top s_k},$$

with an initially given symmetric, positive definite $H_0 \in \mathbb{R}^{n \times n}$,

$$s_k := x_{k+1} - x_k \quad \text{and} \quad y_k := \nabla f(x_{k+1}) - \nabla f(x_k).$$

Then H_{k+1} fulfills the secant equation

$$H_{k+1} s_k = y_k$$

and

$$H_{k+1} \text{ symmetric, positive definite} \quad \Leftrightarrow \quad s_k^\top y_k > 0. \tag{5.6}$$

Furthermore, the inverse H_{k+1}^{-1} can be computed by an update rule, too. Defining $B_k := H_k^{-1}$ the update rule for the inverse is given by

$$B_{k+1} = \left(I - \frac{s_k y_k^\top}{s_k^\top y_k} \right) B_k \left(I - \frac{y_k s_k^\top}{s_k^\top y_k} \right) + \frac{s_k s_k^\top}{s_k^\top y_k}.$$

Thus, the BFGS method reads

- Choose an initial guess $u_0 \in \mathbb{R}^{Nd}$

- Choose initial $B_0 \in \mathbb{R}^{Nd \times Nd}$

- for $k = 0, 1, 2, \ldots$ do

    1. $d_k \leftarrow -B_k \nabla J(u_k)$

    2. Choose a step length $\lambda_k > 0$ such that $J(u_k + \lambda_k d_k) < J(u_k)$

    3. $u_{k+1} \leftarrow u_k + \lambda_k d_k$

    4. $s_k \leftarrow u_{k+1} - u_k$

    5. $y_k \leftarrow \nabla J(u_{k+1}) - \nabla J(u_k)$

    6. $B_{k+1} \leftarrow \left(I - \dfrac{s_k y_k^\top}{s_k^\top y_k}\right) B_k \left(I - \dfrac{y_k s_k^\top}{s_k^\top y_k}\right) + \dfrac{s_k s_k^\top}{s_k^\top y_k}$

    end

Now we have established the BFGS method in general. Nevertheless, some details are left open. Next, we treat the questions how to choose the step length λ_k, how to choose B_0, and what to do if $s_k^\top y_k \leq 0$ and therefore H_{k+1} and B_{k+1}, respectively is no longer positive definite (cf. equation 5.6). Finally, we consider a limited memory version requiring less storage. We start with the choice of step length.

5.3.1 Armijo Line-Search

For the computation of the step length λ_k we propose the Armijo line-search. Let d be a descent direction at a point u, i.e., $\nabla J(u)^\top d < 0$ and $\gamma \in (0,1)$, $\rho \in (0,1)$ fixed chosen parameter. The *Armijo step length* λ is defined as

$$\lambda = \max_{\ell \in \mathbb{N}_0}\{\rho^\ell \; : \; J(u + \rho^\ell d) \leq J(u) + \rho^\ell \gamma \nabla J(u)^\top d\}.$$

Then, first $\nabla J(u)^\top d < 0$ implies $J(u) + \rho^\ell \gamma \nabla J(u)^\top d < J(u)$ and therefore $J(u + \lambda d) < J(u)$. Hence, the Armijo line-search produces a strictly monotonic sequence of function values $J(x_0), J(x_1), J(x_2), \ldots$. Second, we require the function value $J(u + \lambda d)$ must decrease in proportion to the step length λ.

Another important step length is the *Wolfe-Powell step length*. In addition to decreasing function values we require that the derivative of J decreases into the direction of the descent direction d. Therefore, a Wolfe-Powell step length λ must satisfy the condition from the Armijo step length

$$J(u + \lambda d) \leq J(u) + \lambda \gamma \nabla J(x)^\top d$$

and

$$\nabla J(u + \lambda d)^\top d \geq \tau \nabla J(x)^\top d$$

with arbitrary but fixed parameter $\tau \in [\gamma, 1)$.

However, to compute the Wolfe-Powell step length we additionally have to evaluate the gradient $\nabla J(u + \lambda d)$. To this end, and due to its easy implementation here we propose using the Armijo line-search.

5.3.2 Choice of B_0

The BFGS update rule is recursively defined starting with an initial s.p.d. matrix B_0. A quite simple choice for B_0 is the identity, i.e., we set

$$B_0 = I.$$

Then, in this particular case we obtain the descent direction of the steepest descent method, i.e., $d_0 = -\nabla J(u_0)$ for the first step. Setting B_0 the Hessian $\nabla^2 J(u_0)$ results the descent direction of the Newton method. But, on one hand it is not guaranteed that $\nabla^2 J(u_0)$ is s.p.d. and on the other hand as already mentioned dealing with the Hessian is not applicable for us. Nevertheless we have

$$\nabla^2 J(u) = \nabla^2 D(u) + \alpha \nabla^2 S(u) = \nabla^2 D(u) + \alpha h A.$$

When we choose Dirichlet boundary conditions for the discretization of derivatives, the matrix A is s.p.d. and we set

$$B_0 = (\alpha h)^{-1} A^{-1} \approx \left(\nabla^2 J(u) \right)^{-1}$$

The multiplication with A^{-1} can be computed efficiently with the techniques introduced in sections 4.5.3 and 4.5.4. Practical experiments showed this choice for B_0 clearly outperforms the method obtained from $B_0 = I$.

When we choose Neumann boundary conditions the matrix A and its inverse A^{-1} is no longer positive definite but positive semi-definite. To this end, we might introduce in a slight perturbation with the identity, i.e., we set

$$B_0 = (\varepsilon I + \alpha h A)^{-1}, \quad \varepsilon > 0$$

yielding B_0 is positive definite. As above, we can apply the techniques introduced from section 4.5.3 and 4.5.4 for a fast inversion.

Alternatively, practical experiments showed that using the positive semi-definite so-called *Moore-Penrose pseudo inverse* A^+ for singular A also produces good results. However, we will not discuss the use and computation of the pseudo inverse here. We just mention that it can also be computed efficiently with the presented methods. For more on the pseudo inverse cf. for example [32].

5.3.3 Skipping Updates

In order to keep the matrices B_k positive definite the secant updates $s_k = u_{k+1} - u_k$ and $y_k + \nabla J(u_{k+1}) - \nabla J(u_k)$ must satisfy

$$s_k^\top y_k > 0.$$

If we apply the Wolfe-Powell step length strategy this is automatically fulfilled (except if $\nabla J(u_k) = 0$), since

$$
\begin{aligned}
s_k^\top y_k &= (u_{k+1} - u_k)^\top (\nabla J(u_{k+1}) - \nabla J(u_k)) \\
&= \lambda d_k^\top (\nabla J(u_{k+1}) - \nabla J(u_k)) \\
&\geq \lambda(\tau - 1)d_k^\top \nabla J(u_k) \\
&= \lambda(1 - \tau)\nabla J(u_k)^\top B_k \nabla J(u_k) \\
&> 0.
\end{aligned}
$$

Nevertheless, using the Armijo line-search has the computational advantage of avoiding additional evaluations of the gradient but we cannot guarantee $s_k^\top y_k > 0$. To this end, according to [13] if $s_k^\top y_k \leq 0$ we skip the update of B_k in the k-th iteration.

5.3.4 Limited Memory BFGS

As already mentioned in the beginning of this section, we do not save the matrices B_k but the vectors $s_k \in \mathbb{R}^{Nd}$ and $y_k \in \mathbb{R}^{Nd}$. This requires much less storage than saving a matrix, but the required memory increases with the number of iterations, i.e., after k iterations we have to save $k \times 2 \times Nd$ numbers. Thus, for large registration problems the presented BFGS method becomes still too memory consuming after a few iterations. To this end, we use a limited memory version of the BFGS methods (L-BFGS) introduced by Nocedal in 1980 [44]. Therefore, we keep only a fixed number $L \in \mathbb{N}$ of update vectors. After $k > L$ iterations we use only the updates

$$s_k, s_{k-1}, \ldots, s_{k-L} \quad \text{and} \quad y_k, y_{k-1}, \ldots, y_{k-L}$$

and set

$$B_{k-L} := B_0.$$

Thus, the required amount of memory is limited to $L \cdot 2Nd$. Then after $k > L$ iterations the k-th iterate x_k obtained from the L-BFGS method can be interpreted as the L-th iterate of the (memory unlimited) BFGS method started with initial guess x_{k-L}. However, in particular for large scale problems a reasonable performance of the L-BFGS method has been proven with small values for the number of updates L, e.g., $L = 3, 5, 7$ [38].

	#iterations	α	#bins	relative distance
Diffusive	30	10^2	128	0.2222
Elastic ($\mu = 1$, $\lambda = 0$)	30	10^2	128	0.2346
Curvature	50	10^{-1}	128	0.2124

Table 5.2: Parameters for example I

	#iterations	α	#bins	relative distance
Diffusive	30	$5 \cdot 10^2$	32	0.7020
Elastic ($\mu = 1$, $\lambda = 0$)	30	$5 \cdot 10^2$	32	0.7184
Curvature	100	$5 \cdot 10^{-1}$	32	0.7162

Table 5.3: Parameters for example II

5.4 Numerical Examples

To demonstrate the performance of the L-BFGS method, we provide the same numerical examples as for the PDE based method described in section 4.6. The first example is the registration of an academic image pair. In the second example we apply our optimization method for the registration of a CT and a PET image. The resulting images and transformations of the academic example are displayed in Figure 5.9 and the results for the PET-CT registration are shown in Figure 5.10.

As for the examples for the PDE based methods, the parameter α and the number of bins were chosen manually to compute reasonable alignments yielding smooth transformations. The parameters are listed in Tables 5.2 and 5.3. For the diffusive and elastic registration Dirichlet boundary conditions were imposed and for the experiments with the curvature smoother Neumann boundary conditions were used. Furthermore, the images were represented as cubic B-splines and all derivatives (∇T) were computed exactly.

In all experiments the methods performed well and converged quickly. Figures 5.5 and 5.6 show the decrease of the function values of the objective function J. Recall, for the PDE based approach from the previous chapter we do have an objective function that can be evaluated. Therefore, we introduced the relative distance (4.32). For comparison with the experiments in section 4.6, here we additionally computed the relative distances, too. They are shown in Figures 5.7 and 5.8.

A general observation during the experiments was that the method behaved quite robust w.r.t. to changes in the parameter α and the number of bins.

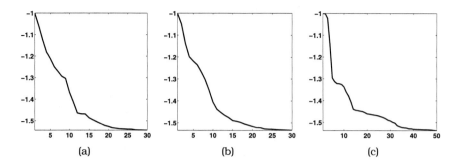

Figure 5.5: Convergence of the BFGS method for example I. (a) Diffusive registration, (b) Elastic registration, (c) Curvature registration

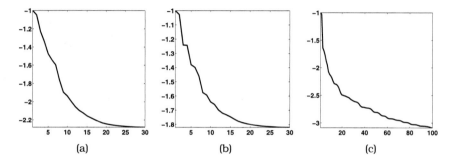

Figure 5.6: Convergence of the BFGS method for example II. (a) Diffusive registration, (b) Elastic registration, (c) Curvature registration

Compared to the results in §4.6, the computed alignments appear visually closer as those for the PDE based method and even the transformations appear to be smoother.

Beside these superior results, the discretize-optimize approach has the charming advantage that we have a (discrete) objective function we can evaluate and use for an automatic parameter tuning. In particular the choice of the parameter τ in the PDE based approach corresponds to step length selection. Here, we are able to use theoretically sound step selection strategies as the Armijo or Wolfe-Powell rule. This is a direct benefit of the discretize-optimize approach and results a quite robust method.

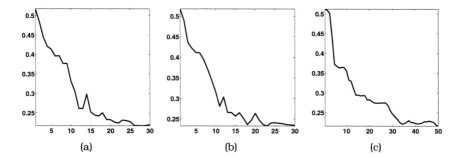

Figure 5.7: Relative distance for example I. (a) Diffusive registration, (b) Elastic registration, (c) Curvature registration

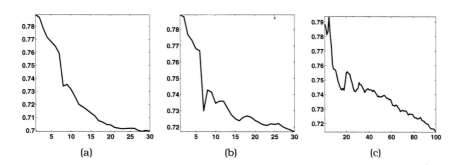

Figure 5.8: Relative distance for example II. (a) Diffusive registration, (b) Elastic registration, (c) Curvature registration

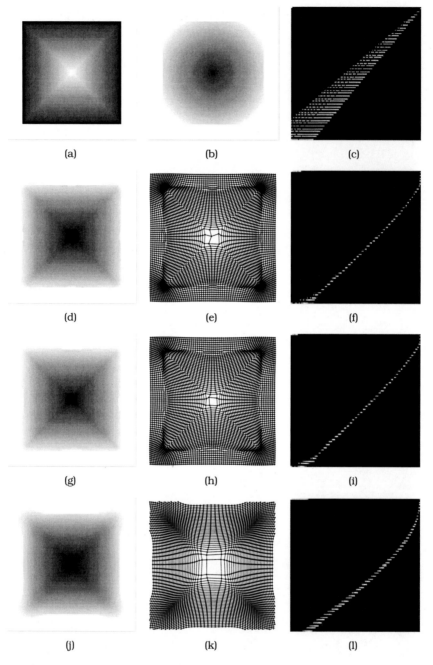

Figure 5.9: (a) Reference, (b) template, (c) joint histogram; (d)–(f) diffusive registration, (g)–(i) elastic registration, (j)–(l) curvature registration; (d), (g), (j) deformed templates; (e), (h), (k) inverse transforms applied to a regular grid; (f), (i), (l) joint histograms of reference and deformed templates

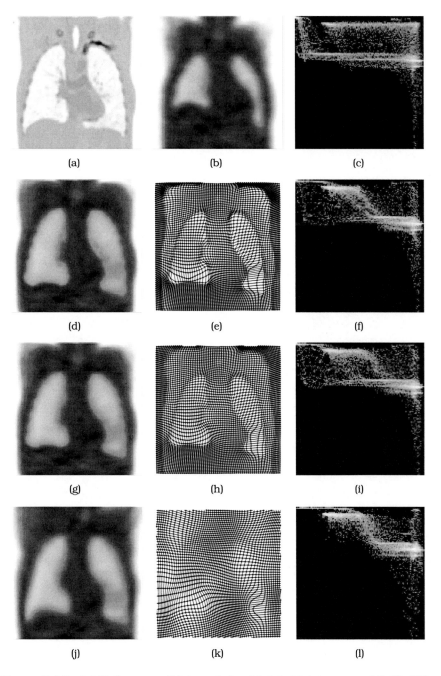

Figure 5.10: (a) Reference, (b) template, (c) joint histogram; (d)–(f) diffusive registration, (g)–(i) elastic registration, (j)–(l) curvature registration; (d), (g), (j) deformed templates; (e), (h), (k) inverse transforms applied to a regular grid; (f), (i), (l) joint histograms of reference and deformed templates

Chapter 6

3D PET-CT Registration

6.1 Introduction

Diagnosis: cancer, one of the most feared and serious diseases in our times. Thanks to intensive research in this field therapies and treatments have been developed which drastically improve the odds of winning the fight against a tumor. Nevertheless, early recognition and treatment is crucial for a successful therapy. Nuclear imaging techniques such as PET (Positron Emission Tomography) or SPECT (Single Photon Emission Computer Tomography) can detect tumors and are major diagnostic tools. However, for a successful and minimal as possible invasive treatment such as radiation therapy or surgery, finding the exact location of a tumor is crucial. Due to the nature of the creation process of nuclear images they do not display detailed anatomical features and have only a low resolution. Therefore, one tries to combine nuclear images with images from other modalities featuring a detailed display of the anatomy. An example for such a modality is CT (Computer Tomography). This is the point where image registration comes into play. To combine for example a PET and a CT image one aims first to align the images and then to find the exact location of "hot spots" displayed in the PET by comparing them with the anatomical structure displayed in the CT image. The alignment is often done manually by rotating and translating an image. State-of-the-art software for computing an automatic alignment usually performs a rigid registration.

In a still ongoing project with the Clinic of Nuclear Medicine of the RWTH Aachen we apply our non-linear registration techniques to register 3D PET and CT images of the chest. The aim of the project is to improve the alignment and thus the localization of "hot spots" in the PET at the CT image.

We start with a short outline of PET and CT imaging and describe the data we are using. Subsequently, we describe the steps taken for the registration and finally present some results.

147

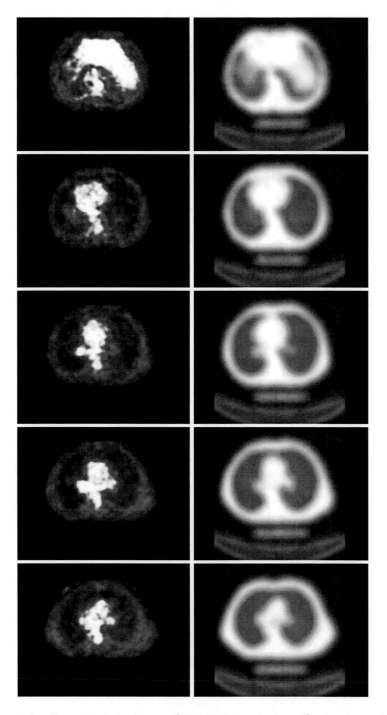

Figure 6.1: Transversal slices of 3D PET emission (first column) and transmission (second column) images

6.2   PET and CT Imaging

PET is a functional imaging technique that has been proven to provide a high sensitivity and specificity to visualize tumors not shown by morphologic imaging modalities such as CT or MRI. For the acquisition of PET images a patient is injected with a tracer. A tracer is a substance that is labeled by a radioactive isotope with a short half-life time (hlt). Typical isotopes are ^{18}F ($\sim$110min hlt), ^{11}C ($\sim$20min hlt), ^{13}N ($\sim$10min hlt), and ^{15}O ($\sim$2min hlt) of the elements fluorine, carbon, nitrogen, and oxygen [65]. Depending on the tracer substance, after injection the tracer concentrates in certain tissues and body regions. In particular, for the detection of malign tumors radioactively labeled glucose ^{18}F fluoro-deoxyglucose (FDG) is widely used. Tumor cells have a high metabolism consuming much glucose such that FDG concentrates at these cells. However, the PET image is then generated from the detected emitted radiation caused by the tracer. Therefore, one also speaks of a PET emission image. Additionally, for the image reconstruction one needs information about the attenuation factor of the tissue that was passed by the detected radiation. To this end, a second radioactive source is placed outside the patient and the attenuation of the radiation when crossing the patient is measured. From this by-product one can also produce a so-called transmission image which is similar to morphologic images such as CT. An example for PET emission and transmission images of the chest is shown in Figure 6.1. PET is a functional imaging modality that does not provide detailed anatomic information. The applied radioactive loads have to be minimal. Due to this, today's spatial resolution in the images is approximately 5mm and cannot be increased without increasing the dose of radiation. Furthermore, acquisition times are rather high. For example, a full body scan takes over an hour and causes many non-linear motion artifacts in the images induced by breathing, heart beat, etc.

To compensate the low resolution and the lack of anatomical details of PET images, CT imaging is used. CT is the oldest 3D imaging technique which was invented by Godfrey Newbold Hounsfield in the early 1970's. Today it is the most common imaging technique in health-care and also used in many other non-medical fields. However, the spatial resolution of modern scanners can be driven to less than 1mm such that CT images display morphological features at a very high level. The data acquisition is based on the attenuation of X-rays of an X-ray source that rotates around the patient. An example for a CT image of the chest is shown in Figure 6.2.

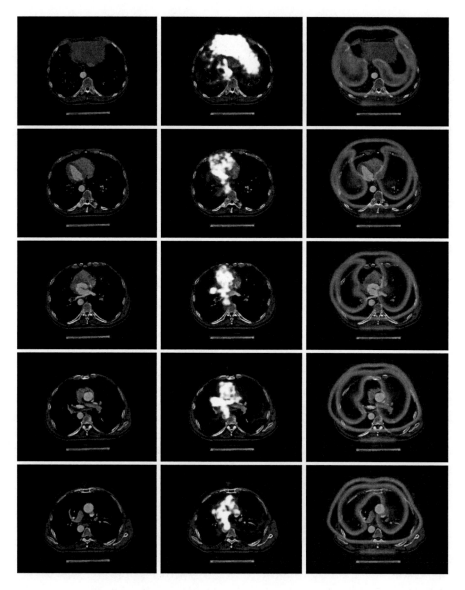

Figure 6.2: Before registration. Transversal slices of a 3D CT image (first column) with fusion of PET emission (second column), and PET transmission (third column)

6.3 Registration Approach

Since PET is a functional and CT a morphological imaging technique, they contain orthogonal information, i.e., PET images display features not contained in a corresponding CT image and vice versa. Therefore, performing a registration of plain CT and PET images will not work in general. To overcome this problem, the idea is to register CT and the PET transmission images. First, recall the transmission image is gained from measuring the attenuation of an out-of-patient radiation source - like a CT image. Second, the transmission and the emission measurement in the PET are made in the same acquisition cycle such that the images automatically share the same geometry. Thus, aligning CT and PET transmission images is equivalent to aligning CT and PET emission images.

Summarizing, we proceed as follows. In the first step we register CT and PET transmission. Thereby, the CT image is the reference and the transmission image is the template. In the second step, we apply the computed transformation to the PET emission image.

6.4 Methods

For the registration of CT and PET transmission images we use the L-BFGS method described in the previous chapter. We favor this approach, because it is more robust w.r.t. the parameters α and σ than the PDE based method and for the advantage of having an automatic step-length selection strategy.

As smoother we chose the curvature smoother. The expected transformation that maps one image onto the other is quite complex, highly nonlinear and yields rather large displacements. Therefore, we made a conservative choice and picked the curvature smoother because it produces the smoothest transformations compared to those obtained from diffusive and elastic registration.

To speed up the computation and making the method more robust we employ a common multi-resolution scheme [24, 40, 57]. To this end we create image pyramids with four levels of $256\times256\times256$, $128\times128\times128$, $64\times64\times64$, and $32\times32\times32$ voxels. At the finest level are the original $256\times256\times256$ images. The next coarser level is obtained by removing high frequencies in the data by a low-pass filtering with a Gaussian and subsequent interpolation to $128\times128\times128$. For the $64\times64\times64$ and $32\times32\times32$ levels we apply the same procedure to the $128\times128\times128$ and $64\times64\times64$ level, respectively. An example for a 2D image pyramid is shown in Figure 6.3.

After creating the image pyramids, we first compute a rigid pre-alignment

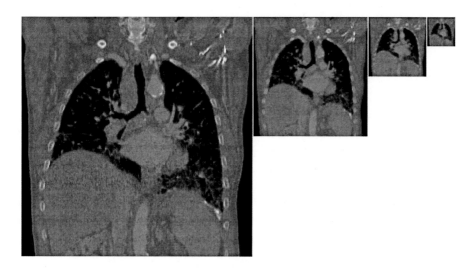

Figure 6.3: 2D image pyramid (256×256, 128×128, 64×64, and 32×32)

of the 32×32×32 images and use the computed transformation as an initial guess for the curvature registration. Then we interpolate it to the 64×64×64 level and use it as initial guess for the registration on this level. Then again, the outcoming transformation is interpolated to 128×128×128 and used as initial guess here. The same holds for the 256×256×256 level.

Finally, we apply the transformation resulting from the last curvature registration that aligns the 256×256×256 CT and PET transmission image to the PET emission image.

A graphical overview on the whole registration procedure is given in Figure 6.4.

6.5 Related Work

In the following, we shortly summarize further state-of-the-art approaches for PET-CT registration using mutual information that were proposed.

In [42] the authors use affine registration and a TPS (Thin Plate Spline) warping based method with few control points for the alignment of CT and PET emission phantom data.

A similar approach was recently proposed in [33], where a rigid registration of whole-body CT and PET emission images is evaluated.

Another promising method is presented in [7]. In a first step, they identify mutual structures of CT and PET emission images using segmentation techniques. Then, the structural information is integrated in a Free-

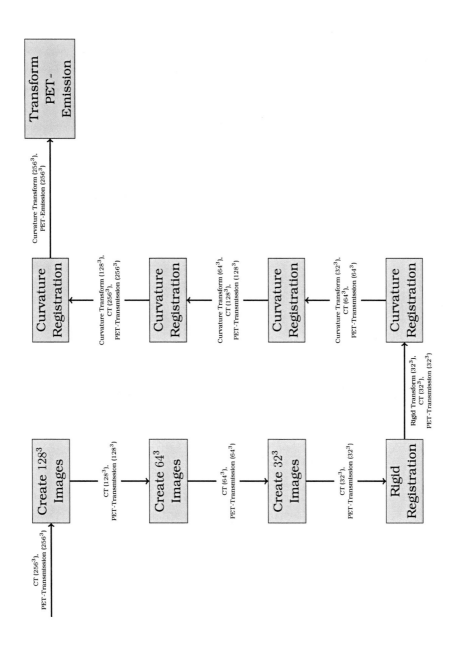

Figure 6.4: Work Flow

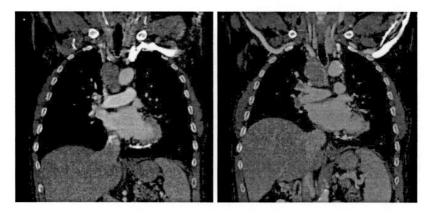

Figure 6.5: CT image of a patient during maximal respiratory inspiration (left) and expiration (right)

Form-Deformation based registration using B-splines.
Similarly, in [52] the authors use segmentation techniques for the registration of whole-body PET and thoracic CT images. In a pre-processing step, they perform a rigid registration of CT and PET transmission images. Then they segment the lungs in the PET transmission image and automatically identify corresponding points. Subsequently, a TPS warping of the images is computed.

Most comparable to us is the approach presented in [40]. The authors propose a Free-Form-Deformation based registration using B-splines to register CT and PET transmission images of the chest. They also employ a multi-level scheme and use a L-BFGS method for the optimization, too.

6.6 Experiments

The data used in the experiments was particularly prepared for the evaluation of our registration method. For the acquisition of the CT and PET chest images positioning aids were used. In each test case (a patient examination) two CT images at maximal respiratory inhalation and exhalation and one PET image were made (cf. Figure 6.5). Both, the CT and PET images were reconstructed with an isotropic voxel size of $1.5 \times 1.5 \times 1.5$ mm^3. The data-volume of each image is $256 \times 256 \times 256 = 16,777,216$ voxels.

In total, we experimented with four test cases. In each case we applied our method for the registration of the CT images at maximal inhalation and exhalation as references and the PET images as templates for several values of the smoothness parameter α that controls the flexibility of the alignment. The values for α differed in five orders of magnitude. Finally, the resulting alignments were visually rated by physicians from the Clinic

Alignment-Error

Structure	Maximal Exhalation	Maximal Inhalation
Tumor	0.0 – 0.5 cm	0.5 – 1.5 cm
Heart	0.0 – 0.5 cm	0.5 – 1.5 cm
Liver	0.0 – 0.5 cm	0.5 – 1.5 cm
Intestine	0.0 – 0.5 cm	0.5 – 1.5 cm
Stomach	0.0 – 0.5 cm	0.5 – 1.5 cm
Backbone	0.0 – 0.5 cm	0.5 – 1.5 cm
Kidney	0.0 – 1.0 cm	1.0 – 1.5 cm
Spleen	0.0 – 1.0 cm	1.0 – 1.5 cm

Table 6.1: Organ specific registration results

of Nuclear Medicine of the RWTH Aachen. Therefore, they individually evaluated the alignment of the following structures:

- Tumor
- Heart
- Liver
- Intestines
- Stomach
- Bone marrow in the backbone
- Spleen
- Kidneys

All experiments were done on a standard desktop computer (AMD Athlon, 2GHz, 2GB RAM, SuSE Linux). The performance of a single registration was in the range of 20–50 min.

6.7 Results

In the experiments we made two general observations. First, the registration of PET and CT images at maximal exhalation gave significantly better results than the registration with the CT images at maximal inhalation. Second, in all test cases the alignment improved with decreasing values for α. The organ specific alignment results for the smallest value of α are presented in Table 6.1. Furthermore, to get a visual impression of the results Figure 6.6 shows the fusion of CT and PET after the registration.

In general, we found good alignments of PET and CT at maximal exhalation which are 0.5 cm accurate where the registration for the CT images at maximal inhalation shows errors up to 1.5 cm. In both cases, the alignments of kidneys and spleens are slightly worse than those of the other organs.

6.8   Conclusions and Future Work

In this project we have demonstrated the proposed registration methods in a non-academic setting. The results suggest that we are generally able to register PET and CT images. The computed alignments are approximately 5mm accurate and are therefore in the range of the spatial resolution of PET imaging.

Furthermore, CT images at maximal exhalation are a substantially better starting point for the registration with PET than those at maximal inhalation. Thus, the methodology plays a very important role and is crucial for a successful registration.

Nevertheless, besides these promising results the trend for better alignments with decreasing values for the smoothness parameter α suggests that a reasonable transformation requires many degrees of freedom and less smoothness. Thus, as a result of these experiments, it seems that in this application curvature registration is too restrictive w.r.t. the smoothness of the computed transformations. To this end, we hope the alignment improves when using the elastic or diffusive smoother yielding less smooth transformations. Therefore, next we want to repeat the experiments using elastic registration. Moreover, we want to confirm our results with the registration of further test cases.

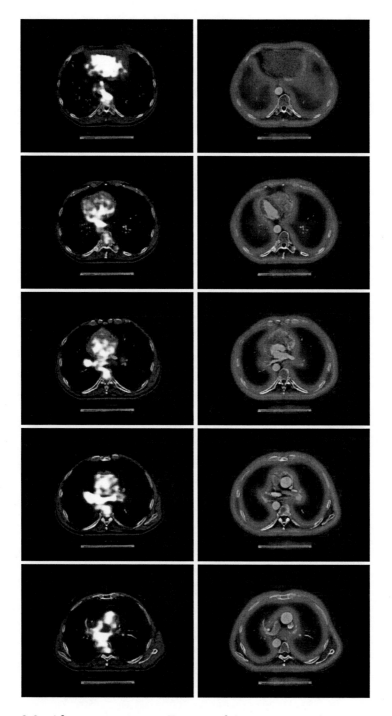

Figure 6.6: After registration. Fusion of CT images with PET emission (first column), and PET transmission images (second column)

Appendix A

Tools from Integration and Measure Theory

Tools from Integration

Definition A.1 (L^p-Norms, L^p-Spaces)

Let $\Omega \subset \mathbb{R}^d$ and $f : \Omega \to \mathbb{R}^k$ be a function. For $1 \leq p < \infty$ we define the L^p-norms

$$\|f\|_{L^p(\Omega)} := \left(\int_\Omega \|f\|^p \, dx \right)^{\frac{1}{p}}$$

and the L^p-spaces

$$L^p(\Omega) := \{ f : \Omega \to \mathbb{R}^k \; : \; \|f\|_{L^p(\Omega)} < \infty \}.$$

Theorem A.2 (Dominated Convergence)

Let $(f_k) \subset L^1(\Omega)$ be a sequence of integrable functions which converge almost everywhere to f. If there exists $F \in L^1(\Omega)$ such that $|f_k| \leq F$ for all k, then the limit f is integrable and

$$\int_\Omega f_k \, dx \quad \to \quad \int_\Omega f \, dx.$$

Proof. For a proof see e.g. [34, §8]. ∎

Theorem A.3 (Divergence Theorem)
Let $\Omega \subset \mathbb{R}^d$ be a bounded open set with piecewise smooth measurable boundary $\partial\Omega$, i.e., $\partial\Omega$ is a C^1-boundary up to a $(d-1)$-dimensional null set and has finite (surface) measure. If $u \in C^1(\overline{\Omega}; \mathbb{R}^d) \cap C(\Omega; \mathbb{R}^d)$ and $\operatorname{div} u \in L^1(\Omega)$ then

$$\int_\Omega \operatorname{div} u \, dx = \int_{\partial\Omega} u \cdot n \, dS$$

where n denotes the outer normal field of $\partial\Omega$.

Proof. For a proof see e.g. [34, §12]. ∎

Tools from Measure Theory

Definition A.4 (Image Measure)
Let $(\Omega, \mathcal{A}, \mu)$ be a measure space and $f : \Omega \to \Omega'$ be a measurable function mapping to a measure space $(\Omega', \mathcal{A}')$. The image measure of μ under f is defined as

$$\mu_f(A') := \mu(f^{-1}(A')) \quad \text{for all } A' \in \mathcal{A}'.$$

Theorem A.5 (General Transformation Rule)
Let $(\Omega, \mathcal{A}, \mu)$ be a measure space, $(\Omega', \mathcal{A}')$ be a measurable space, $f : \Omega \to \Omega'$ $\mathcal{A}$-$\mathcal{A}'$-measurable and $\mu_f : \mathcal{A}' \to \mathbb{R}_+$ be the image measure of μ under f. If g is μ_f-integrable then $g \circ f$ is μ-integrable and

$$\int_{\Omega'} g(y) \, d\mu_f(y) = \int_\Omega g(f(x)) \, d\mu(x).$$

Proof. For a proof see e.g. the books [2, 31, 15]. ∎

Theorem A.6 (Radon-Nikodym)
Let μ and ν be measures on a measure space $(\Omega, \mathcal{A})$ and μ be σ-finite, i.e., there exists a sequence $\{A_n\}_{n\in\mathbb{N}} \subset \mathcal{A}$ such that $\mu(A_n) < \infty$ and $\cup_{n\in N} A_n = \Omega$. Then the measure ν is absolutely μ-continuous if and only if

$$\mu(A) = 0 \quad \Rightarrow \quad \nu(A) = 0 \quad \text{for all } A \in \mathcal{A}.$$

Proof. For a proof see e.g. [2, §17] or [31, §3.18]. ∎

Bibliography

[1] O. Axelsson and V. Barker. *Finite Element Solution of Boundary Value Problems*. Academic Press, New York, 1984.

[2] H. Bauer. *Maß und Intergrationstheorie*. de Gruyter, 1992.

[3] W. L. Briggs, E. v. Henson, and S. F. McGcormick. *A Multigrid Tutorial*. SIAM, 2000.

[4] C. Broit. *Optimal registration of deformed images*. PhD thesis, Department of Computer and Information Science, University of Pensylvania, 1981.

[5] L. G. Brown. A survey of image registration techniques. *ACM Computing Surveys*, 24(4):325–376, 1992.

[6] T. Brox, A. Bruhn, N. Papenberg, J. Weickert, and R. Baum. High accuracy optical flow estimation based on a theory for warping. *Lecture Notes in Computer Science*, 3024:25–36, 2004.

[7] O. Camara, O. Colliot, D. Delso, and I. Bloch. 3D nonlinear PET-CT image registration algorithm with constrained free-form deformations. In *Proc. 3rd IASTED International Conference on Visualization, Imaging, and Image Processing, VIIP 2003*, pages 516–521, September 2003.

[8] E. Chu and A. George. *Inside the FFT Black Box - Serial and Parallel Fast Fourier Transform Algorithms*. CRC Press, 1993.

[9] A. Collignon. *Multi-Modality Medical Image Registration by Maximization of Mutual Information*. PhD thesis, KU Leuven, May 1998.

[10] A. Collignon, F. Maes, P. Vandermeulen, P. Suetens, and G. Marchal. Automated multi-modality image registartion based on information theory. *Information Processing in Medical Imaging*, 1995.

[11] T. M. Cover and J. A. Thomas. *Elements of Information Theory*. John Wiley & Sons, 1991.

[12] E. D. D'Agostino, F. Maes, D. Vandermeulen, and P. Suetens. A viscous fluid model for mutimodal non-rigid image registration using mutual information. In T. Dohi and R. Kikinis, editors, *MICCAI 2002, LNCS 2489*, pages 541–548, 2002.

[13] J. E. Dennis, JR. and R. B. Schnabel. *Numerical Methods for Unconstrained Optimization and Nonlinear Equations*. Prentice Hall, 1983.

[14] G. Egnal. Image registration using mutual information. Technical Report MS-CIS-00-05, Department of Computer and Information Science, University of Pensilvania, Philadelphia, PA, 2000.

[15] J. Elstrod. *Maß und Intergrationstheorie*. Springer, 2005.

[16] L. C. Evans. *Partial Differential Equations*. AMS, 2002.

[17] B. Fischer and J. Modersitzki. Curvature based registration with applications to mr-mammography. In P. S. et al., editor, *Computational Science - ICCS*, volume 2331 of *LNCS*, pages 203–206, 2002.

[18] B. Fischer and J. Modersitzki. Fast diffusion registration. In M. Nashed and O. Scherzer, editors, *Inverse Problems, Image Analysis,and Medical Imaging*, volume 313 of *Contemporary Mathematics*. AMS, 2002.

[19] B. Fischer and J. Modersitzki. Curvature based image registration. *JMIV*, 18(1), 2003.

[20] M. Frigo and S. G. Johnson. FFTW 3.1, 2003.

[21] L. Greengard and J. Strain. The fast gauss transform. *SIAM Journal on Scientific and Statistical Computing*, 12(1):79–94, 1991.

[22] E. Haber and J. Modersitzki. A multigrid method for image registration. Technical Report TR-2004-005-A, Department of Mathematics and Computer Science, Emory University, Atlanta GA 30322, 2004.

[23] E. Haber and J. Modersitzki. Numerical methods for volume preserving image registration. *Inverse Problems*, 20(5):1621–1638, 2004.

[24] E. Haber, J. Modersitzki, and S. Heldmann. Multilevel optimization methods for mutual information registration. Technical Report TR-2004-015-A, Department of Mathematics and Computer Science, Emory University, Atlanta GA 30322, 2005.

[25] W. Hackbusch. *Multi-Grid Methods and Applications*. Springer, 2003.

[26] S. Heldmann, O. Mahne, D. Potts, J. Modersitzki, and B. Fischer. Fast computation of mutual information in a variational image registration approach. In *Bildverarbeitung für die Medizin 2004, Algorithmen - Systeme - Anwendungen" 29. - 30. März 2004, Berlin*, 2004.

[27] S. Henn. A full curvature based algorithm for image registration. *Journal of Mathematical Imaging and Vision*, 24(2):195–208, 2006.

[28] S. Henn and K. Witsch. A multigrid approach for minimizing a non-linear functional for digital image matching. *Computing*, 64(4):339–348, 2000.

[29] S. Henn and K. Witsch. Iterative multigrid regularization techniques for image matching. *SIAM Journal on Scientific Computing*, 23(4):1077–1093, 2002.

[30] G. V. Hermosillo. *Variational Methods for Multimodal Image Matching*. PhD thesis, Universtité de Nice-Sophia Antipolis, 2002.

[31] J. Hoffmann-Jørgensen. *Probability with a View towards Statistics*, volume I and II of *Probability Series*. Chapman & Hall, 1994.

[32] R. A. Horn and C. R. Johnson. *Matrix Analysis*. Cambridge University Press, 1990.

[33] M.-L. Jan, K.-S. Chuang, G.-W. Chen, Y.-C. Ni, S. Chen, C.-H. Chang, J. Wu, T.-W. Lee, and Y.-K. Fu. A three-dimensional registration method for automated fusion of micro PET-CT-SPECT whole-body images. *IEEE Transactions on Medical Imaging*, 24(7):886–893, July 2005.

[34] K. Königsberger. *Analysis 2*. Springer, 1997.

[35] S. Kunis and D. Potts. NFFT, Softwarepackage, C subroutine library. http://www.math.uni-luebeck.de/potts/nfft, 2002.

[36] S. Kunis and D. Potts. NFFT 2.0. Preprint B-04-07, Institute of Mathematics, University of Lübeck, 2004.

[37] S. Kunis, D. Potts, and G. Steidl. Fast gauss transforms with complex parameters using nffts. Preprint A-05-12, Depertment of Mathematics, University of Lübeck, Wallstraße 40, 2005.

[38] D. C. Liu and J. Nocedal. On the limited memory bfgs method for large scale optimization. *Mathematical Programming*, 45:503–528, 1989.

[39] F. Maes, A. Collignon, D. Vandermeulen, G. Marchal, and P. Suetens. Multimodality image registration by maximization of mutual information. *IEEE Trans. Med. Imaging*, 16(2):187–198, 1997.

[40] D. Mattes, D. R. Haynor, H. Vesselle, T. K. Lewellen, and W. Eubank. PET-CT image registration in the chest using free-form deformations. *IEEE Transactions on Medical Imaging*, 22(1):120–128, 2003.

[41] R. J. McEliece. *The Theory of Information and Coding*, volume 86 of *Encyclopedia of Mathematics and its Applications*. Cambridge University Press, 2nd edition, 1977.

[42] C. R. Meyer, J. L. Boes, B. Kim, P. H. Bland, K. R. Zasadny, P. V. Kison, K. Koral, K. A. Frey, R. L. Wahl, and U. E. Aladl. Demonstration of accuracy and clinical versatility of mutual information for automatic multimodality image fusion using affine and thin-plate spline warped geometric deformations. *Medical Image Analysis*, 1(3):195–206, 1997.

[43] J. Modersitzki. *Numerical Methods for Image Registration*. Numerical Mathematics and Scientific Computation. Oxford University Press, 2003.

[44] J. Nocedal. Updating quasi-newton matrices with limited storage. *Mathematics of Computation*, 35(151):773–782, 1980.

[45] J. P. W. Pluim, J. B. A. Maintz, and M. A. Viergever. Mutual information based registration of medical images: a survey. *IEEE Transactions on Medical Imaging*, 22:986–1004, 2003.

[46] D. Potts and G. Steidl. Fast summation at nonequispaced knots by nffts. *SIAM Journal on Scientific Computing*, 24:2013–2037, 2003.

[47] A. Roche. *Recalage d'images médicale par inférence statistique*. PhD thesis, Universtité de Nice-Sophia Antipolis, 2001.

[48] A. Roche, G. Maladain, and N. Ayache. Unifying maximum likelihood approaches in medical image registration. *International Journal of Imaging Systems and Technology: Special Issue on 3D Imaging*, 11(1):71–80, 2000.

[49] T. Rohlfing and C. R. Maurer. Intensity-based non-rigid registration using multilevel free-form deformation with an incompressibility constraint. In W. Niessen and M. Viergever, editors, *MICCAI*, number 2208 in LNCS, pages 111–119. Springer, 2001.

[50] O. Scherzer and J. Weickert. Relation between regularization and diffusion filtering. *Journal of Mathematical Imaging and Visio*, 12(1):43–63, 2000.

[51] B. W. Silverman. *Density Estimation for Statistics and Data Analysis*, volume 26 of *Monographs on Statistics and Applied Probability*. Chapman & Hall, 1986.

[52] P. J. Slomka, D. Dey, C. Przetak, U. E. Aladl, and R. Baum. Automated 3-dimensional registration of stand alone ^{18}F-FDG whole-body PET with CT. *The Journal of Nuclear Medicine*, 44(7):1156–1167, 2003.

[53] W. H. Steeb. *Kronecker Product of Matrices and Applications*. B.I. - Wissenschaftsverlag, 1991.

[54] G. Steidl and M. Tasche. *Schnelle Fouriertransformationen - Theorie und Anwendungen*. Fern Universität-Gesamthochschule Hagen, Fachbereich Mathematik, 1996.

[55] G. Strang. The discrete cosine transform. *SIAM Rev.*, 41(1):135–147, 1999.

[56] P. Thévanez and M. Unser. Spline pyramids for inter-modal image registration using mutual information. In *Proceedings of the SPIE Conference on Mathematical Imaging: Wavelet Applications in Signal and Image Processing V*, volume 3169, pages 236–24, 1997.

[57] P. Thévanez and M. Unser. Optimization of mutual information for mutiresolution image registration. *IEEE Transactions onImage Precessing*, 9(12):2083–2099, December 2000.

[58] P. Thévanez and M. Unser. Stochastic sampling for computing the mutual information of two images. In *Proceedings of the Fifth International Workshop on Sampling Theory and Applications (SampTA'03)*, pages 102–109, 2003.

[59] U. Trottenberg, C. W. Osterlee, and A. Schüller. *MULTIGRID*. Academic Press, 2001.

[60] J. Tsao. Interpolation artifacts in multimodality image registration based on maximization of mutual information. *IEEE Trans. Med. Imaging*, 22(7):854–864, 2003.

[61] C. F. Van Loan. *Computational Frameworks for the Fast Fourier Transform*. SIAM, 1992.

[62] P. A. Viola. *Alignment by Maximization of Mutual Information*. PhD thesis, Massachusetts Intitute of Technology, 1995.

[63] P. A. Viola and W. M. I. Wells. Alignment by maximization fo mutual information. In *5th International Conference on Computer Vision*, 1995.

[64] J. Werner. *Numerische Mathematik 2*. Vieweg, 1992.

[65] Wikipedia.   Positron   emission   tomography.
http://en.wikipedia.org/wiki/Positron_emission_tomography,
February 2006.

[66] L. Zöllei, J. W. I. Fisher, and W. M. I. Wells. A unified statistical and information theoretic framework for multi-modal image registartion. *Information Processing in Medical Imaging (IPMI) 2003, LNCS 2732*, pages 366–377, 2003.

Index